THE GREAT WOOD

The Ancient Forest of Caledon

Jim Crumley

BIRLINN

First published in 2011 by
Birlinn Limited
West Newington House
10 Newington Road
Edinburgh
EH9 1QS

www.birlinn.co.uk

ISBN: 978 1 84158 973 2
eBook ISBN: 978 0 85790 090 6

British Library Cataloguing-in-Publication Data
A catalogue record for this book is available from the British Library

Typeset by Iolaire Typesetting, Newtonmore
Printed and bound by Grafica Veneta, Italy

THE GREAT WOOD

Contents

Acknowledgements   ix
Prologue: A Walk in the Woods on St Andrew's Day xi

 1 Soliloquy by the Fortingall Yew   1
 2 A Lament for the Trees   12
 3 A View of Trees   26
 4 Glen Finglas   46
 5 Sunart   62
 6 Strath Fillan   84
 7 Glen Orchy and Rannoch   101
 8 Rothiemurchus   119
 9 Creag Fiaclach   142
10 Glen Strathfarrar   147
11 The Great Woods   159

Epilogue: Out of the Trees   172

Blessed is the man that walketh not in the counsel of the ungodly, nor standeth in the way of sinners, nor sitteth in the seat of the scornful.

But his delight is in the law of the Lord; and in his law doth he meditate day and night.

And he shall be like a tree planted by the rivers of water that bringeth forth his fruit in his season; his leaf also shall not wither; and whatsoever he doeth shall prosper.

Psalm 1: The Book of Psalms

I do believe in God but I spell it Nature.

Frank Lloyd Wright

Acknowledgements

The author is grateful to the Scottish Arts Council (now Creative Scotland) for a grant that greatly assisted the cause of this book.

Three works by other writers were of particular assistance in my researches: *A History of the Native Woodlands of Scotland* by T.C. Smout, Alan R. MacDonald and Fiona Watson; *A Pleasure in Scottish Trees* by Alistair Scott; *Wild Endeavour* by Don and Bridget MacCaskill. Thanks for permission to quote from all three titles.

A Walk in the Woods on St Andrew's Day

It was St Andrew's Day and I went for a walk in the woods. Autumn had been long and lingering and mellow and mild and extraordinarily beautiful. Waking to the first hard frost, I opened a window, smelled the change, shivered, closed the window. Over breakfast I looked out at foreground trees and wooded hills furred with ermine and blurred by low mist and sang to myself a line of James Taylor: 'The birches were dream-like on account of that frosting . . .' I decided that St Andrew's Day to celebrate trees. I know a place . . .

The sun rises cloudlessly on a whitened land, defeats the mist, but slowly. The land has the sudden pallor of winter. Grasses, smoky orange in October, have bleached to all the shades of straw, and there are not many of those. Bracken has collapsed and darkened to all the shades of unburned toast, but with none of the allure of toast. Trees are battening down and all but bare – oaks hold a few wan, clustery leaves all the shades of cold tea, birches manage a smatter of old gold, palest green willow leaves thin out and whiten, drifting earthwards even without a wind to urge them down. An old hazel wood, crowded out by its own blackened, writhing limbs and looking like so many ancient dancing Masai tribesmen, whispers and spits and bristles at me as I pass, as if it didn't like me to bear witness to its secret dance. Even on a windless day, hazel woods don't believe in stillness.

I fancy that small covey of aspens just heaved a sigh; their heart-quickening autumn flames, the brightest light in any Highland wood, are dowsed beyond recall, and there is no leaf to work their tremulous sorcery. Spring is suddenly far off and unimaginable.

At this moment in the year, the dark green of Scots pine deepens, glows and ennobles. Here is a good one, not particularly tall, perhaps 50 feet, but the trunk is broad and hefty and reddish and round, the bark grooved like tractor tyres, the canopy wide and airy and smoky green. Such a tree is called a granny pine. God knows why.

Few things so ennoble a Highland hillside as a pine-wood spiced with birch and juniper, and every pinewood has its outlying sentinels, stoical stand-alone trees of all but indestructible tenacity. Here is one I have known for 30 years. It is quite alone on its hillside; the pinewood is over there, quarter of a mile away. Fifteen years ago lightning rearranged its handsome presence. A direct hit cleaved its downhill, north-facing limb from the parent trunk, almost but not quite severing it. The limb collapsed into the ground and was quite withered away within two years. But the tree still stands. In fact it thrives to east and west and south. The withered limb still clings to it, and if anything it has strengthened the tree by buttressing it against the steep fall of the hillside to the north.

Seton Gordon, one of the founding fathers of the modern Scottish nature-writing tradition, who spent much of his working life in the Cairngorms and their fringing pine-woods, noted a characteristic of the woods' solitary outliers like this one: 'I do not recall ever having seen a forest outpost or sentinel uprooted by wind; they stand, undismayed, against gusts which send their fellows farther in the forest crashing to the ground.'

He wrote about one such tree in particular in his matchless 1924 book, *The Cairngorm Hills of Scotland*:

. . . a very old Scots fir grows beside the burn. It is called Craobh Tillidh, or the Tree of the Return, and received its name in the old days when a summer population lived at the head of Gleann Einich, and when the stirks and the cows with their calves were driven up a few days before the people themselves went to the shielings. The herdsman accompanied the animals as far as the Tree of the Return. From here the beasts, knowing the road from former summers, were able to continue the journey by themselves and the herdsman returned to Rothiemurchus.

I found the idea of a Tree of the Return an affecting one, even though I never owned a herd of cattle to drive that way. In my own explorations and writings about Gleann

Einich, I first identified another conspicuous solitary pine where I have always paused on my way into the glen, then christened it the Tree of the Beginning. Its story is written down in my own Cairngorms book, *A High and Lonely Place* (Cape, 1990; Whittles, 2000), but my many memories of it are jogged by this wounded pine much nearer to home. Both stand near old tracks, old ways through the hills, old ways into very old pinewoods. Both are recognisable at once and at a distance. The thought occurs suddenly: such trees were always landmarks. Whatever the greatest extent of our native forest may have been, there were always outliers – conspicuous, solitary trees that wandering or settled tribes acknowledged and celebrated; others that stood alone in clearings or at the confluence of rivers or at a meeting of old ways through the land; others that became meeting places, sacred places, churches without walls; others that marked the sites of great events, great lives lived, epic battles, love trysts . . .

Perhaps more than anything else, that single idea – a reverence for individual trees – is the symbolic bond that is still capable of evoking fellow-feeling with those who have walked this way before, not just Seton Gordon's herdsman, but those shadowy folk who walked a much more liberally forested land several thousand years ago.

Like all symbolic bonds, its significance is perhaps overdone. Likewise the notion of that forest that popular imagination sees as something that smothered all but the mountaintops; that was reputedly prowled by slavering tribes of wolves and bears and big cats and cattle called aurochs the size of elephants, and terrifying blue-painted men; that we – or someone who walked this way before us – gave a name to, the Great Wood of Caledon.

So here is a good Scots pine, a wounded veteran of a life-long struggle waged against this irksome climate, and if you linger like me by its tortured limb and rest your back against its still vigorous trunk and look west out across the crowns of the pinewood that soften the bulwark of mountains beyond, or north to where the low sun of St Andrew's Day has made a yellow patchwork of still more mountains, and if you let your seeing eye wander there, you will find other solitary pines to set alongside other ways into the hills, other ways into the fertile history that is enshrined in our notion of the Great Wood.

But ways into the future too, for among those solitary pines are a few that preside over new pinewoods, where something is being put back, where thousands of hand-planted head-high pines, and oaks and aspens on the lower ground, and birches and rowans that have sown themselves uninvited (but always welcome) and alders that shade the riverbanks are returning. It is happening in a small way, but it is happening here and there all across those parts of Scotland where 21st-century science tells us that native woodland once grew. I fancy I sense the weary approval of these few stalwart old trees that have borne witness to our recent past, the last two or three centuries, say, centuries that have been characterised perhaps more than any other of human occupation by crimes against the landscape, against trees, against all nature.

So I had decided that St Andrew's Day was a good day to celebrate trees. I spent it in and around the pinewood and walked out in the late afternoon, pausing again by the wounded sentinel pine, the tree of a new beginning. An old friend of mine who had been a forester near here was much in my mind. His name was Don MacCaskill. He died about ten years before I wrote this, and his philosophy seems to be finding its way more and more into my books. On the last page of *Wild Endeavour*, the first book he wrote with his wife Bridget, there is this:

An area of woodland is the ultimate habitat and must be saved and expanded at all costs. The forest is a complex entity, a mirror of nature, and has the desired structure wherein a balance can be achieved. The people of this country, in whose history there is little of forest background, sometimes think that trees are an intrusion into the familiar scene of smooth bare hillside, or cotton grass bog. They do not understand that the heather or bracken-clad hill are arrested habitats which, if left to nature, could well become forest once again. A good deal of Scotland is a peat desert with severe erosion problems and a very restricted species structure. Trees provide a blanket to shelter the vulnerable soil, arrest erosion, provide oxygen and give a home to a wide range of species, be it bird, animal, plant or insect. A forest is a complex world, not easily destroyed. Without trees it is an over-simplified world, and one that is very vulnerable.

So it was in the spirit of finding something in my own lifetime to act as a kind of bridge to link what has gone, what is, and what can be, that I decided to go for a walk in the woods, to celebrate those trees that were the landmarks of my country for thousands of years, and to do so at a time when we are beginning to grasp the essential truth of Don MacCaskill's philosophy.

Here and there and all across the face of the land, an old mist is rising in the face of something hopeful and enlightening, and the way ahead is slowly growing green again, and the Great Wood stirs from a long and ominous slumber and begins to throw new shadows in the resurrected sun.

Before the Resurrected Sun

Trees inhale in gasps,
the coldness rasps
in the oldness of their limbs
and sticks in the craws of ravens.
And still the thaw withholds,
and still the red deer pause,
keeping the wood's edge near.

So it was when ice held sway
before the resurrected sun
arose and ice gave way,
and then
the ever deeper print of men.

CHAPTER ONE

Soliloquy by the Fortingall Yew

A high stone wall embellished with thick iron railings and a hefty iron gate with a discouraging padlock stands in the corner of a Perthshire churchyard. It is a strange place for a prison. The solitary inmate is the oldest living thing in Europe, and just possibly in the whole world. But for the last 150 years or so it has languished in solitary confinement, apparently the consequence of too many Victorian souvenir hunters bearing saws and pruning shears. In the last 20 or so of the 150 it has been further demeaned by our own generation's obsession with walk-this-way visitor interpretation that tells you what to think and feel and assumes you have a mental age of five. So my first impression was that something extraordinary and unique had been diminished by small minds.

I began by just watching through the railings, which, without a key to the gate or a substantial hacksaw or enough gelignite to blow the lock, is the only way you and I can watch and wonder. I was reminded at once of a snow leopard I saw long ago in Edinburgh Zoo, the magnificent made miserable. It looked miserable and it evoked miserableness in the watchers beyond the bars. Its plight offered a two-way exchange of miserableness. I was instantly and irredeemably against all zoos forever. Unlike the snow leopard, the creature imprisoned in this churchyard has no eyes for its incarceration to deaden, does not pace out its daily ritual along a bare path the width of itself. But see! There is a circle of inches-high posts inside its stone-and-iron cage, and that circle is the width of the creature's trunk in its prime – 56 feet. Before it began to succumb it may have reached 60 feet in girth.

When it was younger and unincarcerated it was free to wander the wild wood by way of its several thousand generations of offspring; these were borne out into the world by winds and by those countless tribes of the wild-wood that wittingly and unwittingly carried and scattered the seeds of trees, and these would include bird, bear, boar, wolf, man.

It is argued of course that its incarceration has spared it from death by a thousand cuts from a thousand pairs of pruning shears and from tourism's worst excesses. But coming to stand and stare in something of the spirit of pilgrimage like so many before me, I was confronted by this zoo creature – dishevelled, decrepit and wretched, cowering in its dark corner – a creature as brought down by its circumstances as a snow leopard that has forgotten how to be exquisite or how to wear the tribal finery that is its birthright.

I came on an April day of cheerful winds and on-off suns, the air fast and fizzing like swifts. I dutifully took the approved path to the tree with its tourist-trade messages carved into the flagstones, and felt my eager mood evaporate. The air barely stirred in the tree's compound, and sunlight was rationed there to a few bright scraps the size of fallen leaves that had squeezed through the bars to tease ancient limbs.

And science has done this. It has made a freak show of perhaps the oldest living thing, taking and tending cuttings to sustain the oldest known pedigree, but at the expense of the spirit and the will of the living tree. I want to liberate it, to bulldoze the prison and let in the wind, the sun, the rain, the snow, the frost, to let it be a tree again for what remains of its life and let it take its chance with the rest of the world as it did for all of those unknown thousands of years before the jailers arrived.

I watched a group of loudly chattering visitors plod round the path, reading the panels and trying to decipher the flagstones. A woman of about my own age detached herself, skipped tourism's preamble, and came over to where I was standing. She fretted by the railings. She had a camera in her hand. I sympathised:

'It's not easy to photograph, is it?'

'Oh, I don't care about that,' she said, 'I'd just like to touch it.'

A small green plaque was there and several other plaques on the outside wall offering visitors sundry scraps of information ranging from mildly interesting to worse than useless. None of these was quite as useless as the small green plaque. It has been planted *inside* the compound, but positioned so that it can be read through the prison bars. It was planted in 2002 to mark the Queen's Jubilee, so that it might proclaim the Fortingall Yew as one of Fifty Great British Trees, and adding insult to that particular injury, it declares that it is sponsored by the National Grid.

So it has come to this. The single living reference point to the so-called Great Wood of Caledon, the only living thing that can claim with all honesty the right to say 'I was there', has become an advertising hoarding for the National Grid. Whatever the Great Wood was, this is its one miraculously alive survivor. By what preposterous arrogance does anyone get to sponsor a plaque to tell the world that such a tree is 'great'? And what on earth has it got to do with the Queen?

If the National Grid really feels the need to pay tribute to the greatness of trees let it sponsor the means to return to nature arguably the greatest of all known trees by virtue of its age alone. Pull down the walls and let it breathe, let it bow to winds and warm to suns. And yes, let the woman standing next to me touch it and take home with her whatever it is that she would take from a moment of intimacy with a living organism whose very life is so far beyond mortal comprehension.

So, I told myself, understand that first: it *lives*. The cowering, shadowed creature still lives. Then what?

Then consider the potential of yew trees. This one is very probably upwards of 5,000 years old and very possibly nearer 10,000 than 5,000. No one knows, not even to the

nearest thousand years, how old it is. You need the heartwood to acquire such knowledge, but old yew trees are
hollow and where the heartwood once grew there is now a
gap almost 60 feet wide. What does grow there now is a pair
of unlovely thin trunks on the outskirts of the tree that was.

The Great Yew stands in the village of Fortingall near
the mouth of Glen Lyon and a few miles west of the
Highland Perthshire town of Aberfeldy. Whether or not
it is the oldest living thing in Europe or the whole world, it
will certainly be the oldest living thing I will ever rest my
eyes on, you too if happenstance or pilgrimage cause you
to pass this way. So when you consider the potential of yew
trees it is almost disappointing how quickly you become
nonchalant about a thousand years here or there, how casual
you become with your 'probabilities' and your 'possibilities' and how interchangeable they suddenly seem. Science
pronounces its best guesses, then qualifies them with an
embarrassed shrug, then disagrees with itself in the face of
some new theory. Science does not always like to own up to
what it does not know. Specifically, it does not like to be
defeated by a single tree. This is something from which I
draw considerable comfort.

There again, whether it is 5,000 years old or 8,000 or
9,000 hardly matters, for it is in any case a number far in
excess of anything you and I can comprehend when applied
to the lifespan of a single organism. Even 500 is nebulous
enough, and 500 years old is the point at which (in the event
that it survives that long in the first place) a yew tree begins
to grow again. It's a sort of second childhood. So 5,000 years
old should be like being born again ten times, living

biological proof of reincarnation. But in order to reincarnate you must first die, and the Fortingall Yew has not died; except that somewhere along the line reincarnation gives way to deterioration and the thing begins to die, bit by bit. Science's determination that it should apparently be kept alive forever by means of taking cuttings and replanting them seems to me to suggest a lack of faith in the immortality of trees. Shame. Yet also, bit by bit, it begins to live again, even from its ever-depleting resources. There is a green canopy above the mysterious darkness, above the stone walls and railings. New sprigs of green still reach eagerly towards the south-facing gate, for that way lies sunlight, and every living tree, whatever its predicament, strives towards the sun. Seeds still grow in their due season, fussy seeds like tiny acorns life-belted by a red berry-like flesh, somewhere between 5,000 and 10,000 harvests, an incomprehensible fecundity.

Here then is an invitation to consider a different definition of time from the one we routinely use that recognises days and hours and minutes, and when it considers years at all it thinks mostly in terms of human lifetimes. The Fortingall Yew invites you to dare to consider the possibility of immortality. Fred Hageneder, an American tree expert, wrote in his 2001 book, *The Spirit of Trees*: 'While the heartwood inside the hollowing trunk slowly rots away, sheaths of new growth encase the old dead wood to strengthen and protect it. Thus yew renews itself from the outside in . . . A yew that appears to be a hollow, decaying wreck is often at the beginning of its self-regeneration process. Yew can resurrect itself from complete decay.

There is no biological reason for a yew tree to die – it can virtually live forever.'

But immortality in trees is not a new idea, even among nature writers, and especially among the pioneers of the American nature writing tradition. Some cite species other than yew. John Muir had the Sierra juniper in mind when he wrote: 'Surely the most enduring of all tree mountaineers, it never seems to die a natural death, or even to fall after it has been killed. If protected from accidents, it would perhaps be immortal.' And Donald Culross Peattie wrote of the giant sequoias: 'Those who know the species best maintain that it never dies of disease or senility. If it survives the predators of its infancy and the hazard of fire in youth, then only a bolt from heaven can end its centuries of life.' Or not; witness the Scots pine I passed on my St Andrew's Day walk.

We know now that Scots pines can live for three or four hundred years, and that in places like the Cairngorms and the hills in and around Glen Affric there are populations of pines whose lineage can be traced directly back to some of the earliest known forests in the land. We can walk in such forest remnants now and something reaches out to us that is outwith the scope of plantation forests in which great age is routinely missing. Ralph Waldo Emerson wrote that 'in the woods, too, a man casts off his years, as the snake his slough . . . In the woods, we return to reason and faith.'

Hmm, maybe. It is said that John Knox liked to preach under yew trees; if he did, he just rose in my estimation. But that was a tradition already old when he would have taken it up. It may also have been a nod to a still older and undatable

Celtic tradition (but certainly one that was already old when Christianity was born) that associated yew trees with sacred sites.

All the writers quoted above were moved by trees much younger than the Fortingall Yew. The more you consider it, the more extraordinary its survival seems, and the less sceptical you become about the possibility of an immortal tree. Consider, for example, the discovery in 1991 of the Ice Man in a frozen glacier in the Alps on the border between Italy and Austria. His preserved Stone Age body was accompanied by an equally well preserved axe handle and a bow stave, both made of yew. Apart from anything else, there was never a more telling demonstration of yew wood's capacity to resist rot and damp. We already knew about that, but what about ice? If the wood can emerge intact from millennia of incarceration in a tomb of ice, are there yew trees out there that survived the ice age, whether as seeds or even living trees? Was this at Fortingall one of them? Just how old is the idea of an immortal tree?

Yew bark, foliage and seeds (but not their berry-like coverings) are known to be poisonous, and the potent darkness within the cloistered space created by the drooping curtains of a mature tree's foliage creates uneasiness in a susceptible human mind. In fact it is unarguable that the yew, of all tree species, has resonated uneasily in the human mind for as long as the two species have cohabited in the same landscape. It still does. Here, for example, in a brief interview for the book *Flora Celtica* (Birlinn, 2003), is Peebles-based bagpipe maker Julian Goodacre on the subject of his raw materials: 'Different woods evoke different feelings in me. I love yew. We use a lot of yew. It has strange associations though, and is a curious wood. Like the tree, the timber imparts a sense of foreboding. It contains poisons and mysteries, and I feel uneasy about breathing its dust.'

Is that sense of foreboding what John Knox had in mind? Was his liking for a congregation held in the round embrace of a yew's drooping canopy a psychological trick? If he preached 'I am the Light' in the pervasive gloom and the uneasiness of the yew's dust, was the congregation only too willing to lean towards the Light he had to offer?

However far back you go, you find yew trees in high places. *A History of the Native Woodlands of Scotland* by Smout, MacDonald and Watson (Edinburgh University Press, 2005) notes: 'Irish Brehon Law of the eighth century classified trees into four classes with seven species in each: the classes were termed nobles, commoners, lower orders and slaves, reflecting both the Gaelic sense of hierarchy and the economic importance of each group . . .' Needless to say, the yew was a noble.

And there is another potent hint from within the Gaelic sense of hierarchy that appears to elevate the yew's sacred symbolism. In the Gaelic language's 17-letter alphabet each letter is represented by a tree, and the letter I is *Iogh*, standing for *Iubhar*, the yew tree. And in the wake of St Columba's decision to set up camp on Iona, the island acquired a one-letter name – I, and the name lives on in the island's single conspicuous hilltop, Dun I. It may have nothing to do with the yew tree, of course, and Iona is hardly the first place you would think of in terms of woodland, sacred or otherwise. Or it may have everything to do with it.

9

Much earlier sacred sites than Iona were beside land-mark yew trees, and it is at least plausible that before sacred buildings, the yew trees themselves were the 'churches', such is their capacity to create a small curtained space where people could gather and shelter and their rituals were screened from view. In time, burial mounds and other pagan structures were built beside the sacred trees, then the Christian ones were built on the site of the pagan ones. Whether your Christianity leaned towards Columba or Knox or something in between, the yew tree was the easily recognisable symbol, and its earthy enclosure was – still is – as moving and affecting a space as a wee kirk or the chapel of a great cathedral.

So when you pause in your 21st-century travels by the Fortingall Yew and feel it manipulate your mind and invite you to consider the power of trees, perhaps it is presenting you with the key to the padlock. Perhaps it is asking you to consider the historic place of trees in the landscape, and, in the context of Highland Scotland, that means grappling with the concept of the Great Wood.

Ah, if only I might interrogate the Fortingall Yew, I could untangle a few strands of truth from the thorny understorey of the myth-makers, let in light to illuminate history's shadows. But the yew is a husk of the tree that was, and husks are poor conversationalists. The two ragged trunks and sundry twisting tendrils crouch by the compound's furthest wall. There are too few crumbs of sunny comfort to relieve the elegiac gloom in which the yew passes its days.

I wanted to write something that gave me a clearer idea of the Great Wood, one that an exploration of the landscape might sustain, so I started here with this singular tree and the idea in my head of its un-walled stance on a south-facing slope before the church-builders set to, a tree with an extraordinary girth (strange: no one ever seems to have taken the trouble to measure its height), and a reputation as old as the hills for hundreds of human generations among whom it was both a landmark and a symbol. For whatever the scope of the Great Wood may have been, great trees within it would be known and revered by all the tribes of the land.

It seems to me that the way to honour this of all trees would be to make space around it instead of walls, and to plant a new wood around the church and the burial ground and the neighbouring land, using the species that would have been the yew's natural companions, then let nature make of it what it will. Because that is the way that we – and nature – have treated woodland forever: we have worked with what's there and the way that it grows, we have adapted it to suit our purpose and nature has adapted what we have adapted. And sometimes we have introduced trees from beyond Scotland to try and improve on the native mix, and nature has adapted these too, and worked with them, and so trees move back and forward across the face of the land as they have always done.

11

CHAPTER TWO

A Lament for the Trees

The first nature writers sang the praises of the land and sea and sky and all that moved there – fish, deer, cattle, horse, lynx, boar, beaver, bear, badger, skylark, swallow, swan, eagle, wolf. They sang the praises of mountains where their various gods made love and war, fire and ice. They sang the praises of trees. Nothing they made is signed. We don't know who they were. Whatever they wrote is wreathed about by the cold, corrosive patina of millennial dust. But then as now (and whenever 'then' may have been and whoever 'they' may have been), their themes were the preoccupations of their age, and trees provided them with shelter, boats and oars, bows and arrows and spears, ploughs, fruit, nuts, and the makings of fire that changed everything. Trees mattered to them. Some were landmarks, some sacred, some both. The first 'churches' were trees: nave, chapel, cloister and spire in one ready-made kit. An early bard might have written:

The trees of the Great Wood, they were as clustered and prolific as stars in the midwinter sky, they had ascended to the mountains' shoulders and they would soon have conquered the stars too, softening the profile of Venus herself, smothering the Red Star with the green of pines, had not the mountains suddenly begun to shrink to accommodate the capricious demands of the gods, their volcanic fire, their glacial ice; begun to shrug the trees from their lowered shoulders with every new upheaval of granite, gabbro, sandstone, gneiss. The gods had grown anxious at the forest's advance, fearful that the trees might smother their realms, and then how would they see the earth beneath, how watch over and manipulate the land and the sea and those creatures they hade made in their own image? How find the North Star that made sense of all their heavens if their view was impaired by a thick screen of branches?

So before the trees grew so dense that they shut out the view of the Milky Way and made a darkness of the aurora, the gods shrank the mountains with their fire and ice, and the trees slithered away downhill. The gods commanded the grazing tribes – the aurochs, the deer, the boar, the horse, the goat, the sheep, the beaver, to put the trees in their place. They equipped the beaver with the instinct both to graze and hew, the people with the instinct to hew and burn. All that not only put the trees in their place, it also made their place smaller and smaller and lower and lower down the mountainsides so that they dwindled the way a flood tide abates to a low ebb. For of course the gods overdid it, the way gods do when they get the bit between their teeth, leaving the impoverishment of lone and level sands and salted mud, and all of that was as naked and cold as Venus in the winter sky.

Those first nature writers had begun by praising the trees for all their gifts. Then they saw how it was between the trees and the gods, and they began a new work, and that was a lament for the trees.

Woodland That Was Not

Scent the distilled whisky of the land.
Scan the sheep-shorn glen.
Toast the woodland that was not.
Drink:

To every willow
that never wept with the joy of being.

To every silver birch
that never found its crock of gold
at summer's rainbow's end.

To every rowan
that never raised a green banner over an eagle's throne
and to every eagle eyrie never built
and every eaglet
that never fledged and never flew
from a rowan-bright nursery.

To every hazel, oak and alder
that never shadowed the burn
and every trout and salmon
that never lingered in pools never shaded.

To every songbird
that never pierced each silent May Day dawn
and never lived to die in the fast clutch
of every sparrowhawk
never weaned in nests that never leaned
by tall pine trunks that never grew
in the woodland that was not.

To every tree-creepering, wood-peckering, owl-
    hootering thing
that never clawed bark that never wrapped
all the ungrown wood,
and every roe and stoat,
badger and bat,
squirrel and wildcat,
four-legged this and that,
that never stepped into clearings
all across the whole unwooded glen.

To every woodland moth and mite and moss
and tree-thirled lichen,
a health to you wherever you prospered.
It was not here
in the glen grown barren as a hollow tree.

*

So the greatest days of the Great Wood began to unravel,
began to evolve into a series of disconnected smaller woods,
although the final shape of that land we know today as
Scotland was still a couple of thousand years away; it was no

more than a few millennia ago that the seas rose and smothered the last vestiges of our land bridge to mainland Europe. Something like those smaller woods still survives, or at least their impoverished direct descendants do. A frail but crucially unbroken thread of ancestry remains in place, perhaps 5,000 years old (and a handful of isolated and even more hoary remnants of six, seven, eight thousand years), and they are as haphazard and thin on the ground as pearls in a stream. But it is from such as these that old Scotia's grandeur really does spring, and a 21st-century nature writer must cast about among them for an admittedly inadequate sense of their great strongholds. A lament for the trees is his task too.

We have become accustomed these last 200 years or so to talk up the heyday of the trees in a particular form of words, 'The Great Wood of Caledon'. In our minds it is almost exclusively associated with the Highlands, and with Scots pines. And it was a dark and fearfully impenetrable shroud. Even science is still apt to fall back on the careless notion that the Great Wood covered all of the Highlands. And because two foresters named H. M. Steven and A. Carlisle wrote in a landmark book *The Native Pinewoods of Scotland*, published in 1959, that 'to stand in them is to feel the past' and that Scotland has only one per cent of its native woods left, and because these sentiments appear to have extra resonance in our sound-bite-hungry century, you hear them again and again from the mouths of people who might have been expected to know better.

I have stood in almost all of Scotland's pinewoods, some of them many times. To stand in them today is to feel anything but the past, other than in the sense of a lament for that which was lost by climate change throughout their history, and more recently by the actions of many generations of our ancestors that brought much of what was left to its knees, or even its ankles, obliterating much of the wildlife of the treed landscape in the process. Rather, to stand in them is to feel loss, to feel the fragility of what remains, to fear for their future. And yes, to marvel and to admire the power and the beauty and the rightness of native trees in a native setting. I think, for example, of Glen Strathfarrar in late October, where the beauty comes at you in waves and puts an ache in your heart. And you can drive past the road end, as most of us do, and never know it's there, and that massive indifference is fatal.

As to the idea that we have one per cent of our native forest left standing, it may be true that there was a time in the 10,000 years since the ice age when this land of ours was home to extensive forests rather than disconnected woods. But that state of affairs was as false a representation of trees in our land as what we have now, for that forest had grown from the nothing that emerged from the ice age. The first tentative and thinly scattered woods that became a forest over 5,000 years then began to thin again as more and more of nature's tribes thrived there (including our own tribe, of course) and the climate suffered wholly natural convulsions.

All we can say today with any confidence is that in many parts of Highland Scotland, there are far fewer trees growing now than there were when the Great Wood was at its most extensive. If you insist on paring down the comparison to native trees alone, if percentages are more important to you than a healthy, diverse and widespread forest, then you can use numbers to paint a miserable picture. But if you do want to make comparisons like that, why not take the Great Wood back to its ethnically pure origins of eight or nine thousand years ago, when the first pines, willows, birches, and alders – not forgetting at least one yew – put down the first roots of the Great Wood. Then see how far forward in time you can travel before the first hazel, aspen, oak or rowan, say, washed up from some distant shore, or was dropped onto fertile ground by migrating bird or beast or man or woman. For every tree species that ever grew on Scottish soil grew somewhere else first. And whether it was introduced by what we think of as natural means, or by our own species (which we think of as unnatural, as if we were an aberration instead of just one more of nature's species that does some things by accident and some by design), is less important than what we have since chosen to do with the ones we introduced by design, less important than whether or not we give nature its head to make the most of the opportunities at its disposal.

Even as I write this, the Forestry Commission has set aside a small Scottish acreage as a plantation of giant redwoods, a Californian native, against the day when climate change may render it extinct in its own country. It is a slow-growing tree and reaches improbable age and height. I have never seen redwoods in their native land but I have stood often in that small cluster of half a dozen redwoods in Edinburgh's Royal Botanic Garden, and consider it one of the most moving tree places I know.

If the Commission wants to embellish my native landscape with a few hundred redwoods they will get no argument from me. From the first, trees have come and gone from the Great Wood, and they always will. And over there, on the hillside across the glen from the window where I write, the lowering October sunlight makes a daring show of lit gold among the larches that race up the hills in tongues of flame among the solid green of the spruces. That wood, commercially planted, is home to its own population of red deer, badgers, red squirrels, red foxes, among much else, and if you insist that the larch is an alien incomer that should be driven from the land along with the Sitka spruce, then you might as well insist that only west and east winds should blow in Highland Scotland, and that those of the south and north have no place here.

There is also this. Just because the seedbed of the Great Wood has gone from much of the *visible* landscape does not mean it has gone entirely. It is true of the 21st-century Loch Lomond and the Trossachs National Park where I live, and it is true of every historic homeland of trees. Wherever we lift the inhibiting pressure of our lifestyle and leave a space large or small, watch the trees push in. In the very changed circumstances of our own time, the trees still present themselves, still swarm slowly across the land wherever they are left to determine their own journeys, their stances, their particular strongholds. Where humankind has made its many inroads, determining which trees should be permitted to grow and where, nature makes what it can of the changed circumstances. Half a mile from my desk is a fenced-off area of perhaps three acres. It had been a fragment of a large commercial plantation, but now it is privately owned. The owner had it clear-felled of spruce trees then simply left it to its own devices. Ten years later, not ten thousand, it is a self-sown birch wood – nature making what it can of the changed circumstances. Rowans gather along its edges and even prance precariously along its drystane dyke, skinny roots clawing thin sustenance from thick moss and fissures between stones.

On the barest of mountainsides where regimes of sheep and deer and burning have held sway for 300 years now, find a loch and look at its small ungrazed islands, find Scots pines there, rowans wedged among rocks, a low fuzz of juniper, a birch or two – always a birch or two. Likewise along the hillsides, look where the steep-sided Highland burns clatter down through dark gullies that discourage grazers, find them thickened by oaks, birches, alders, a copse of vivid autumn-yellow aspens.

Here are the ruins of an eighteenth-century inn and its outbuildings, owned – they say – by a brother of Rob Roy MacGregor, and where Major Caulfield fashioned a stretch of one of his eighteenth-century military roads from the track between the ruins, the whole footnote to our turbulent history made almost illegible by huge oaks, sycamores, and ash trees that no one planted. Follow the military road north up into Glenogle, where a railway line once traversed the hillside, and the ill-starred Dr Beeching used a landslide as the excuse to terminate that particular endeavour. It is as if the hill above the military road wears a belt about its waist, a belt of birches, of oaks, willows, ash, rowans. And where a stone bridge spans one of those steep burns, the old trees of

the gully and the younger ones of the railway have collided
and made a new wood on the open hillside. The very rocks
of the landslide have been colonised by trees where they
have come to rest in an arrangement too haphazard and
chancy for grazing animals. But badgers have found the
combination of trees and rocks to their liking and moved in.
These things tell you that here, for many, many centuries,
trees and hillsides were accustomed to each other.

Everywhere you look beyond the old and established
woods and the new and established commercial plantations,
you find abundant evidence that the treed landscape is
nature's preference. The deer forest (and whoever came up
with that form of words to describe such an enforced
treelessness had a pitiable sense of irony) and the grouse
moor with its meticulously tended heather are not inven-
tions of the natural order but rather perversions of it. Red
deer are woodland animals, both historically and biologi-
cally; everyone who knows anything at all about deer also
knows that they prosper in woodland and that the red deer
of the Highland hillside, for example, are stunted creatures
by comparison. My old friend Don MacCaskill put it thus:

In the present day, there are red deer who have their
territories and live out their lives entirely on the open hill;
there are those whose territories are permanently within the
forest; and there are those who use the forest verges for
shelter, but who move out onto the hill to feed, or to rut
during the season. But, in bygone ages, the red deer used to
be entirely a creature of the forest . . . splendid stags with
massive heads, and sleek, sturdy hinds, were to be found

roaming the sparsely populated country. They were larger beasts than we know today. Woodland clearings and pine forest provided good shelter and feeding, and natural predators, wolf and lynx in addition to the fox and the eagle, would have killed off the weak and sickly, thus ensuring that only the fittest survived to continue the species.

As the old forests began gradually to disappear, over the centuries the red deer was increasingly deprived of its natural habitat. It was forced to become a creature of the bare hills and glens, and there it survived, but at a price. As a pine, planted at an unnaturally high altitude in poor soil grows, but is small in size producing only inferior needles, so the red deer on the high hills and barren wastes continued to exist, but as a smaller animal, the stags growing only inferior antlers.

So Landseer's Monarch of the Glen is nothing more than an admittedly well-executed piece of Victorian propaganda. And he would have known as well as anyone that in the forest the relationship between people and deer is utterly changed, whatever the motive of the people, whether painter or shooter (all those damned trees) or nature writer. Don MacCaskill once told me: 'A forest is not a forest without deer.' When he told me this he was chief forester of Strathyre Forest and his employer was the Forestry Commission. Needless to say he was not a typical head forester and he was often at odds with the Commission's sometimes institutionalised way of working. But in the matter of deer and forests he was right, and both red and roe deer love pinewoods, old and new.

I was remembering an old autumn, a long day out on Beinn a' Bhuird, ice on its high lochans and a fur of snow on the ice. The wind bit like crampons. The massive nature of the landscape (*massif* is a particularly good word for the Cairngorms), the toughness of that landscape, that environment, that season (early winter in every characteristic bar the date on the calendar), that wind . . . all that had been a brutal taskmaster, brutal and relentless. The pinewood under the mountain, when I reached it late in the afternoon, was spring water in a desert. At once the wind was elsewhere, the temperature was almost comfortable, the light softened, and adversity was overwhelmed by a sense of peace. I walked as deep into the trees as the wood allowed (for it is a small survivor, beleaguered by deer) then sat down with my back to a pine trunk that stood ankle deep in blaeberries and heather. I leaned my head back against inch-thick bark (the vertical landscape of treecreepers), closed my eyes and breathed in the scent, felt the stillness on my wind-fired face. I may have dozed briefly.

I opened my eyes when I heard the stamp of many feet; heard the jangling clatter of antlers; heard the anthemic throb of the open throat of the red deer rut. I have watched the rut a hundred times on open hillsides, in corries, on island shores where stags thrash their antlers in seaweed, but until that moment I had never seen it or even contemplated it in a mountain woodland. There is this essential difference: that it announces itself while it is still invisible, so that its approaching tensions infect every creature in its path, which, as it happened, suddenly included me.

That awareness was instant and acute. But I did not dare

move for fear of what I might miss. I learned long ago that the best hours in nature's company are achieved when I allow it to come to me, and while putting that philosophy into practice has given me my share of heart-in-mouth moments, I have learned to trust it utterly and fear plays no part in it. I was wearing the forest shades of dull green and grey and I told myself: be a bit of tree trunk.

The hinds came first, entering my field of vision at a placid trot that rather devalued the furore of the overture. They slowed to a walk and paused by the river to drink. I counted 14. The sounds of battling stags shifted around left and right but offstage and behind my back. The knowledge that I now sat between the stags and the hinds laid an edgy *frisson* on the moment. It is rare in any endeavour to be so vividly alive, with every sense demanding to observe.

The first stag I saw was going backwards, head down and twitchy. He passed me on my left two yards away. Not much more than the same distance on my right, the reason for his twitchiness announced himself with the rawest, most awful noise I have ever heard this side of a snow avalanche. My head jerked that way of its own volition, despite my advice to myself that I should be a bit of a tree. There stood the master.

He was peat-blackened, barrel-chested, thick-necked, high-headed, wide-antlered. And he stank. He stepped past me, advanced four strides towards the young stag that was preoccupied with one straying hind. At the sound of the master stag, the young animal turned to face whatever was coming his way, took a blow in the neck that drew blood, shied like a horse, swerved away, conceded. He retreated

24

down an alley in the trees, where, as it happened, the stray hind stood side-on and indifferent. But with all options suddenly removed, she turned and ran before him so that in his hour of defeat, and quite involuntarily, he had gleaned a harem of one.

The master stag slowed, bellowed a rebuke at the retreating young stag, turned his attention to the hinds, tried to outflank them so that they could not travel back the way they had come. But he found his manoeuvre constrained by trees and they went anyway. He followed, as stags must in the rut, with only 13 hinds at his disposal.

I imagine the whole episode had taken no more than two minutes. I sat still in stunned and sunny silence and in a kind of privileged shock. I registered only the lingering stench of the stag and the bark-spitting quest of a hunting treecreeper somewhere above my head.

*

I remembered the day as one that had elevated the place of trees in the landscape to something more profoundly elemental. From that day, that old autumn, I began to revere wild woods. And Don MacCaskill had just told me, 'A forest is not a forest without deer.'

CHAPTER THREE

A View of Trees

I have a view of trees walking
the earth on elephant feet,
a march so millennium-slow
it escapes our notice

that the next tree
has just stepped purposefully
into the vacant spoor of the last,
and that the gap

between their repositioned trunks
is but a single stride.
Stumbling blocks are these:
rock, wind, thin Highland soil,

and ourselves, the tribe
that fells elephants.

The march began in the West
when the Great Ice ended,
shrank and wended away
inland, uphill, until

the last grey glacier failed. Behind,
each withering yard was dogged
by the slow jog of the trees,
freed from seeds

that outlived the Ice, or bore
ashore in the warming sea,
or flew with birds seduced
by the temptress sun.

And in the remembered heat
trees walked the earth on elephant feet.

I have a view of trees from the window where I write.
Without standing up or turning my head I can see birch,
ash, rowan, willow, oak, alder, hawthorn, Scots pine, larch,
hillsides full of Sitka spruce, and a distant cluster of big
conifers in someone's garden that may be Douglas firs. I
know from countless explorations of the glen that these
trees share the landscape to one extent or another with
aspen, hazel, juniper, yew, beech, sycamore, maple, black-
thorn, hawthorn and holly. There may well be others and I
could make more of the list if I was a bit more knowledge-
able about which of the ten species of native willow are out
there. The list of things I don't know about trees is a
thousand times longer than the list of things I know.

Balquhidder Glen lies in the north-east of Loch Lomond and the Trossachs National Park, a heartland glen halfway between Scotland's east and west coasts, and a well-treed landscape by the standards of today's Highland glens. Of course, not all the trees are old or native. The biggest landowner inside the national park is also the biggest landowner in Scotland, the Forestry Commission. The favourite tool of its trade is the Sitka spruce, which is more or less routinely derided by the rest of us. It is the tree we all love to hate. The only thing we hate more in our landscape is the midge. In my view of trees, there are a great many Sitka spruces, and in any assessment of the Great Wood of Caledon the Sitka spruce has no place, right?

Wrong.

Because if the Great Wood of Caledon is anything at all, it is an evolving continuum. Here and there across the Highlands the native forest still lives in relict communities: Rothiemurchus, Glen Affric, Torridon, Knoydart, Sleat, Rannoch, Creag Meagaidh, Glen Orchy, the Trossachs, Sunart, Seil, to name but a handful. And although these are all depleted, much tampered with, and their communities mostly isolated from each other, they amount collectively to a kind of skeletal outline, the bones of what a Great Wood might have been. Throughout its life, and beginning in the painfully slow years of its protracted birth in the aftermath of the Great Ice around 10,000 years ago, the Great Wood has evolved, has ebbed and flowed in tides of trees. The twenty-first century finds it at a low ebb, but not its first low ebb by any means, and there are signs that its fortunes may be turning again after the prolonged lethargy of slack water.

The Great Wood began, once the Great Ice relented, with a handful of species, the pioneer tribes: juniper and birch and willow and Scots pine among the earliest, and, as I have suggested, quite possibly yew, given its capacity to survive almost everything. But other species came. They came by air on oceanic winds, they came by sea washed ashore from countries beyond the ice, and they came by the land bridge that still linked Scotland and Europe. They came with human settlers and would-be conquerors. So at what point do we decide that a tree species is an alien? Who makes our tree immigration policy?

The Sitka spruce is a native of Alaska, not Scotland, but it is a natural in Scotland, perfectly suited to our latitude and climate, and demonstrably it thrives here. It may be a late arrival in the context of the Great Wood, but the Great Wood itself was a late arrival in a landscape whose oldest rocks are 4,000 million years old. What we have chosen to do with the Sitka spruce in Scotland is another matter, but I believe that the Great Wood can be a phenomenon with a future as well as a past, in which case the Sitka spruce *will* be a part of it, and as things stand, an essential part. If nature finds it easy to live with here – and it does – why shouldn't we?

So the view of the trees from my window differs at least by degrees from what we are inclined to think of as the landscape of the historic Great Wood. It is missing a few tree species, and every now and again a hillside is clear-felled in a quite unnatural way of which neither nature nor I approve. It is also missing quite a few mammal and bird species, but most of these are things within our control, and if we choose to put them back, if we choose to work the land with a philosophy more hand in glove with nature, we can. Now, as we agonise over the fate of our warming planet, now that conservation and the environment are higher than ever on our political agenda, now there is no better time for us to decide if putting these species back is what we want to do for the landscape. For that matter, shouldn't we also make an educated guess at what nature would have us do for the landscape? And if we do decide to put these species back, there could be no more idealistic declaration of our intent than putting them back into a reconvened Great Wood.

All of which begs the obvious question: what *was* the Great Wood of Caledon?

For that matter, who first called it great (for it cannot have been the National Grid) and how reliable was their assessment? The reputation of the Great Wood that has been handed down to us has all the hallmarks of the handed-down reputation of the Highland wolf (and that is a grotesque caricature, a hybrid of legend, superstition, lies and ignorance, all of it finally drenched in an over-cooked and over-seasoned broth of Victorian invention that too many people swallowed whole, and for that matter too many still do)*. Likewise, we have been led to believe that the Great Wood was impassable, a huge tree ghetto of innumerable terrors which lay in wait to prey on terrified travellers. As with the wolf, that reputation evolved and grew bloated over millennia even before the Victorians got their hands on it.

---

* See this book's predecessor, *The Lost Wolf* (Birlinn, 2010)

The concept of a Great Wood as something other than just a well-treed Highland landscape would appear at first glance to have been invented by the Romans, and its christening a casual one, apparently based on the word-of-mouth testimony of Roman soldiers written down by Tacitus and formalised by Ptolemy back home in Rome in the second century AD. It was Ptolemy's use of the phrase Caledonia Silva that survived to appear on maps produced centuries later in mainland Europe, in particular one in 1513 and another in 1654, but the area of the wood appears to have doubled on the second map. That in turn may well have been a consequence of the influential Scottish historian Hector Boece's ever-so-slightly over-the-top rendering of the Great Wood in 1527. *The History of the Native Woods of Scotland* says of Boece: 'For him, the Roman wood had stretched north from Stirling and covered Menteith, Strathearn, Atholl and Lochaber and was full of white bulls with "crisp and curland mane, like feirs lionis". There is no evidence in classical sources for this embroidery, but in any case the wood was located firmly in the past by Boece.'

A third map in 1708, however, showed scattered, smaller woods rather than a single Great Wood, the most significant of them apparently rooted around Glen Orchy. And that seems to confirm that by then the Great Wood *was* history. Besides, we know from 21st-century carbon dating of the bleached remains of old scraps of trees preserved in peat in landscapes like Rannoch Moor that the *Caledonia Silva* as chronicled by the Romans was in decline for 3,000 years before they got here.

So it was all a bit casual, the vaguest of beginnings for

something that has become such a fixture in one of the stoorier corners of Scottish history. The history of the Scottish people has been a constant preoccupation of academics for many centuries. Geology explains a lot about the history of the rocks beneath our feet. But the history of the landscape on the surface remains our great unknown, as devoid of the crucial heartwood as a 5,000-year-old yew tree.

It is very possible that the Romans exaggerated the difficulties which the *Caledonia Silva* posed because they failed to penetrate it significantly, because they were at the extremity of their empire and their comfort zone, because they didn't much care for what they saw looking north from the ramparts of the Antonine Wall or their fort at Callander. The Pass of Leny 1,500 years before the A84 was built would not have been without its ferocious aspect. Nor did the Romans much care for what their galleys reported back when they went prospecting up the west coast. So, for domestic consumption back home, they prepared a portrait of impenetrable mountain forests populated by fearsome creatures including white bulls like fierce lions and wee painted people. It's not as if they were unfamiliar with mountain forests and wolves and lynx and brown bear, for they had them all at home, but something failed them here that had not failed them before, and they felt the need to blacken the reputation of the landscape before they finally faded away whence they had come. Centuries later, that fearful vision of a Great Wood which they had taken home with them returned to us through civilised European channels rather than an invading army. It returned to haunt us, and we seem to have been willing to believe in it, as we have been willing to believe in the storytelling tradition that also reached us from Europe and blackened the reputation of the Great Wood's most influential, non-human creature, the wolf. What is unclear is why native knowledge – the Picts' knowledge for example – of both the wolf and the Great Wood was so eagerly discarded by much later generations of Scots in favour of the received wisdom of strangers.

A twentieth-century American writer, Hal Borland, observed that 'when man's earliest ancestors achieved reason and dreams they found the makings of tomorrow on the forest floor, saw the future's shape in the long shadows of the woodland . . .' and that was just as true here too. When the first hunter-gatherers washed up on Scotland's post-ice-age shores around nine or ten thousand years ago, with only a few hundred years of primitive woodland growth in place since the retreat of the ice, there was no Great Wood, simply because there had not been time for one to grow.

The first coastal explorers, then, found a young wood, and the human population grew slowly and matured as the wood became great. The people and the Great Wood evolved together, but the wood evolved faster. There were millions of trees by the time there were thousands of people. In time, the people would maintain – or create – clearings for settlements and grazing. They managed small areas of the native forest almost from the outset. It is how people and trees co-exist. Besides, in a northern hemisphere wilderness, which is what the first people encountered, nature works with varied densities of trees as well as varied species; with every age of tree, with grassland, with heath, with wetland, with bog, and with a natural treeline often halfway up the mountainsides. It also works with grazing animals and their predators, and that dynamic relationship also insists on open woods with many wide clearings. So, for that matter, do big winds, and no matter how much the climate warmed and cooled and warmed again, Scotland was the mountainous northern half, first of a peninsula, then an island in a northern ocean, and wide open to winds from the Atlantic west and the Arctic north.

That combination of factors does not beget dense jungle. The Great Wood was never impassable. So when the Romans decided to pull back from this, their outermost frontier, it was almost certainly not because the landscape defeated them, but rather because they had finally overreached themselves. Sooner or later, all empire-builders overreach themselves. They left behind the unconquered Picts, for example, who seem to have managed fine in their *Caledonia Silva*, and who treated the wolf as a revered hunter and a teacher, much as the native North Americans did before the advent of the white man. Its portrait appears on their carved stones and their jewellery among other revered creatures and sacred objects. And the Scots, when they came from Ireland with their Christianity-peddling emissaries, were undeterred by wolves, by forests, or by pretty well anything at all. The Scots liked what they found in Caledonia, including what remained of its *Silva*, and they stayed. Fifteen hundred years later, they're still here.

One aspect of Boece's assessment of the Great Wood is worth considering – the extent of its realm that 'covered Menteith, Strathearn, Atholl and Lochaber . . .'. These are all names whose notional boundaries have contracted over the centuries. Today, for example, Menteith is a loch mistakenly called a lake, a village on its shore, and a small range of low hills. In Boece's day it was the generic name for much of what now constitutes the Loch Lomond and the Trossachs National Park. Atholl might well have encompassed much of the Cairngorms, and in terms of the Great Wood was indistinguishable from the extensive pinewood tracts of neighbouring Badenoch. Boece was Aberdeen-based, so perhaps he can be forgiven for writing off everything west of the Great Glen. The mountainous bulk of the Monadhliath west of the Cairngorms might have been such a physical and psychological barrier that it never occurred to him that the Great Wood might extend west of the Great Glen. The Monadhliath never seems to have figured in any roll call of the wooded Highlands and the same is true of several other parts of the country where mountains dominate utterly. But west of the Great Glen, the parallel east-west glens of Strathfarrar, Cannich, Affric, Morriston, Garry and Arkaig all still have their pinewood souvenirs; here was a substantial stronghold of the Great Wood, one that probably extended south and blended into the oakwoods of Morvern and Sunart in coastal Argyll. The pinewood glens are in varying stages of decline and recovery, but these oakwoods are carefully managed and conserved in the twenty-first century, to a telling and quite beautiful effect. Fragments of coastal and island woodlands from Skye to Mull and as far south as Seil also fit the notion of a Great Wood that was as prolific north and west of the Great Glen as it was south and east of it.

All those western woods, whose every characteristic was shaped by the Atlantic Ocean, would greet many of the first settlers, and given the ubiquitous occurrence of hazelnuts among the unearthed remains of the earliest human settlements, it may be that those tough, wind-and-salt-sculpted Atlantic hazel woods that still cling to our coasts are the direct descendants of the very first seedbed of the Great Wood. Gavin Maxwell wrote in *Ring of Bright Water* that 'man must still, for security, look long at some portion of the earth as it was before he tampered with it'. In the context of the Great Wood, perhaps only the Atlantic hazel woods still offer us the opportunity to do just that. On the other hand, if man now tampers with the notion of the Great Wood to good effect, the day will dawn when his descendants can look at some portion of the earth as it was after he tampered with it because he will have put back his idea of how it was to begin with. And if he lets nature be his tampering guide, his descendants will have cause to be grateful.

The inexplicable omission in Boece's assessment of the extent of the Great Wood is arguably the stronghold at the heart of it all, from Rannoch to Glen Orchy. The all-but-treeless Rannoch Moor we know today coughs up bone-white tree remnants thousands of years old all across its mighty girth, and the Black Wood of Rannoch is still a vigorous pinewood with a back-to-the-ice-age pedigree. There are remnants at Dalrigh in Strath Fillan near

Tyndrum, at Loch Tulla west of Rannoch and in Glen Lyon. And in Glen Orchy the demonstrable tree fertility of today's glen confirms the claims made by several historians of a long woodland pedigree. The riverbank is thick with willows and alder. Old stands of pines, oak, aspen and hazel cluster randomly on the lower slopes. Rowan, birch and aspen grow profusely wherever they can grab a foothold, and there are miles and miles of flourishing commercial plantation.

So I have a view of trees in my mind's eye, disorderly invasions of trees slow-marching across the land; first the empty land left by the ice, then, as the climate grew benevolent, more and more trees in perpetual and irresistible conquest. But this was a benevolent conquest that took a land that had been made desolate by the ice and smothered it with life, with colour, birds, beasts, plants and a million or two lowlier life-forms, until at last a far-flung and airy spread-eagle of forest, a Great Wood, healed the ice-wounded Highlands. There is no end to the daring or the march of the trees.

*

The first snow of winter came with the first day of December. It did not linger on the lower forested slopes for more than a couple of days although it camped on the hills for a fortnight, the snowline advancing down and retreating uphill in wraiths of cloud as temperatures rose and fell. The morning after that first big snowfall was grey and still as a graveyard, the next the snow had dwindled, patchy and drippy by noon, and all the length of the glen the clouds and mist rose and fell and collided and overlapped, a slow pageant of mostly monochrome tones. Spaces kept appearing high up and low down, spaces with trees in them, trees set against snow and broken shards of cloud, outlines softened and sharpened by the restlessness of the lowering clouds, the swithering mist. In such midwinter hours, the Great Wood of Caledon is easy to believe in. It's easy to believe that it's there now, that if you were to walk up in the dusk to where woods cling to high crags in the moonlight, you would find wolf tracks in the snow; believe too that if you raised your voice in imperfect imitation of a howling wolf – your own lone wolf howl – the pack would answer from above, from the midst of the cloud-and-mist wraiths, their upper edges moon-silvered. The small flocks of tourists who gather in every month of the year round the grave of Rob Roy MacGregor in Balquhidder kirkyard would dine out for the rest of their lives on the memory of that edgy thrill.

All afternoon I had walked in the snow. Sometimes the cloud and the mist conspired to close the land down to fifty square yards, drifting in among the ranks of spruce, larch, birch and scattered coveys of rowans, so that it seemed as if the trees themselves were absorbed in a slow and silent dance, advancing and retreating, now solitary, now a swirling host. And all the while my own footsteps in the snow orchestrated the dance, for nothing else sounded, no other rhythm, no other sound of wind or rain or moving water lent its pulse to the day. Sometimes I followed deer tracks, once the bounding spoor of a pine marten, once the dead-straight contour of a fox that had carefully placed its right hind foot in the pad of its right forefoot, the left hind foot in the pad of the left forefoot. A wolf, the greatest of all snow travellers, does the same thing. That way, the animal's footfall only has to break through the snow crust once for every two steps. But mostly I broke my own trail, pausing often to snatch views above or below, and ahead or astern, whenever the grey dancers drifted apart and holes opened, holes that accommodated endless variations on a theme of trees.

Balquhidder is a great deal more wooded than it was 40 years ago. I have two OS maps of the area, one from about 1965, and also a newly minted one. The most conspicuous difference is the amount of green. Much of it is plantation forest, of course, but native trees, notably birch and rowan and oak and alder rush into every unplanted acre. The wide, flat floodplain of the River Balvaig is bordered by thickly wooded banks. Its northern edge, which was once divided into crofting strips, is no longer worked. Instead, birch and alder wade in the winter floods the locals call 'Loch Occasional', and whooper swans glide among the trunks. You give this glen half a chance and it sprouts trees as readily as an untended garden beckons to weeds. Unchecked, nature could make a small Great Wood of the glen in 20 years.

But Balquhidder is southern Highlands, a dozen miles from the Highland edge at Callander. Its climate is moderate by Highland standards. Cross the watershed of Glen Ogle going north and you arrive in Glen Dochart, a great trough of a glen wide open to the assaults of east and west winds, and the beginning of a northerly march into ever more spartan lands. It's not that trees don't grow in more northern glens, it's just that they need more encouragement and fewer inhibiting factors like sheep and deer and absentee landlords who reckon the worth of their estates in the number of stags and the acreage of manicured grouse moor they carry.

Looking down on wooded mountainsides from above the treeline when the mountains are white and happed in fleeces of dense grey that occasionally rearrange their folds to bare a hill shoulder, a gully, a crag, a stand of Scots pine, a clearing among trees where a red deer stag is suddenly revealed and staring at you so that you wonder how much longer he has been aware of you than you have of him . . . all that was the landscape of my afternoon as the winter daylight faded from its dull zenith to a gathering gloom in about three hours. A raven called from the bosom of a cloud, a two-syllable shard of sudden sound that echoed from a rockface. The bird called again, again the echo. I wondered if it could hear the echo and was fooled by it. Then I wondered if it knew about the echo in a certain part of the glen and understood it and was playing with it. But it did not call again and the glen lapsed back into silence, a silence somehow deepened by its intervention.

I looked at it all, a long and lingering look, felt the landscape gather round me and claim me for its own, and I felt a helpless love for that moment in that landscape, in that circumstance. I spoke a verse of Kathleen Raine out loud:

> Though all must pass
> There will never be a time
> When I will not have been
> This here and now
> Of clouds moving
> Across a still night sky.

It is a landscape I know quite well, not well like the stag or the raven knows it, but well enough to navigate its advancing dusk with confidence, intimate enough with its rise and fall and darkening curves where the wood reached up with a beckoning embrace, or down to the distant road home. I stepped into the trees and new snow followed me down. There seemed then to be no end to the trees, as if they were there forever, and undefeatable.

The thought lodged in my mind like a hook in the mouth of a salmon: is that what unsettled the Romans? Such a conspiracy of wildness on such a midwinter night? On top of everything else? They were not nature writers in love with the landscape after all, just young soldiers too far from home.

*

I have another view of trees. If I get up from the desk and go outside, turn my back on Balquhidder Glen and face east, the world is closed in by a long, steep hillside that wears the characteristic working clothes of the Forestry Commission, what the writer David Craig called 'the dark trees of government'. The principal shade is spruce green, enlivened here and there by larches – the springiest of pale green in spring, the fieriest of gold and amber in autumn, brittle-bare in winter. We tend to like the larch, yet it is as 'alien' as the Sitka spruce, although you could argue that 400 years is quite long enough to be regarded as native. I so argue. And so for that matter is the 200 years or so that the Sitka spruce has been growing here.

But this is an uninspiring view of trees. It's too slabby, the treeline is too inflexible; the effect is essentially con-trived, designed on a piece of paper then implemented, unnatural. A first glance takes in a mile of such unnatural-ness, but it actually extends almost to the very southern-most edge of the Highlands ten miles away, and a couple of miles further north to Loch Earn, a rampart of the Strathyre Forest. Here and there along that rampart there is – there always is – a new area of clear-felled forest, a new area of replanted clear-fell in its cheerful infancy, so that over decades the forest is a shifting patchwork of young and middle-aged trees. It is tempting to argue that such a forest has no old trees at all, for the Commission does not permit old age, but that is not quite true. It is true that there are no old trees of the harvested species, but if you were to study a planting map of the forest you would see that here and there throughout its length there are compact knots of native trees, mostly Scots pine or oak, the unlikeliest of survivals, protected oases of age. I draw a bead on one such from the back door and find its tousy fringe with difficulty, for it breaks no skyline, and its own shade of green is not so different from the massed and straight-backed ranks of the spruces – a little bluer, perhaps, but only in the bright light of midsummer, and that is not now. Now is still winter. Now is new snow, a great deal of new snow that still hangs heavily on the trees. I find the Scots pines I am looking for because they widen at the top and the spruces narrow, so they wear their snow differently from the spruces. They do everything differently from the spruces.

The forest road passes the door heading south, climbing, then it hairpins up the east slope and doubles back to a long and almost level north-south terrace. It is a surprisingly open thoroughfare. Verges and ditches and scrubby corners accommodate an equally surprising variety of growing things in spring and summer, from early purple orchids, butterwort, wild hyacinths and bluebells to honeysuckle, brambles, blaeberries, wild strawberries, foxgloves and rowans. Birch trees try to get going everywhere, and here and there it is permitted and here and there it is not, and I have never been able to reckon the why and the wherefore. In winter the ditches run with water or thicken with snow or ice or both, and on the road and its banks and ditches you can read the passage of fox and deer and pine marten and red squirrel and sometimes an otter trying to cross watersheds; wildcat if you're very, very lucky. One day, perhaps, a wandering wolf, for wolves can – and do – live in commercial forests.

You walk by huge larch trees, and these too have been spared from felling the last few times this hillside was cleared of its spruces. They lay an orange carpet on the road in October and November; snow and winter winds banish it. Then you reach the pines and all is briefly changed. The pines stand on a steep hillside below the forest road, in a cluster. Clustered trees that widen at the top make a canopy, so the light changes as well as the shape of the trees. Sunlight and shadow suddenly have something to say. And a canopy, if it is old enough and undisturbed enough and not overgrazed, creates conditions in which the trees' preferred understorey can thrive.

The tight-packed planting regime of commercial spruce obliterates understorey. Penetrate ten yards into such a plantation and the floor of the forest is simply bare under a mat of spruce needles, and dark as the trees strive for the distant sun. But this canopied acre of Scots pine is knee-deep in heather and blaeberry and crowberry, among others. It is as different from everything around it as a coral island is from the sea. The one prerequisite for such a habitat is the passage of time. These trees are old, or at least this pinewood is old. The Commission is a newcomer to this hillside – about 60 years. People and deer and sheep would long since have reduced the old woods to scraps by the time it moved in, and the brave-new-world zeal of the early Commission planters doubtless accounted for a few more precious but commercially inconvenient souvenirs of the wildwood.

But Don MacCaskill became chief forester of Strathyre Forest and Don was a friend to the old woods. The example he set and the legacy he left for his successors have been a thrown lifebelt for woods like that small pinewood, woods within woods where people can pause and admire and look and learn, and where, if the Commission and the Scottish government were ever to unite in common cause behind a single benevolent vision, this very forest, this very national park, might blaze a trail whose destination was something like a reconvened Great Wood. Almost all the ingredients we would need are already in place. It's just a matter of adjusting the balance. The wolf is still missing though.

CHAPTER FOUR

Glen Finglas

Let me return to that other view of trees, the view from the window where I write. Over that hill to the south-west, there is an old way through quiet hills beyond. It climbs at last to a watershed whence it descends to the true heart of the Trossachs and the village of Brig o' Turk. To reach that watershed today, I must walk through hills made bare by the familiar conspiracy of too many deer and sheep and too much human indifference. The riverbanks and a few crags and gullies beyond the reach of the grazing hordes betray the once-wooded nature of the place, for birches, alders and rowans still gather wherever they find a safe haven. But the hills are browbeaten bare. Yet as I cross the watershed and begin to descend the sunny, south-facing slopes beyond, the land begins almost eerily to transform. For there has begun the work of a project called 'Return of the Forest'. Welcome to Glen Finglas.

I like watersheds, like how they organise the landscape, build bridges between mountains and command rivers where to flow. I like how they lift deer, fox, wildcat, otter and me from one glen to another (they also used to lift bear, boar, lynx, wolf). I like how they give my wandering footsteps cause to pause. They are edgy, exposed places. Breast a watershed – what's next?

But the slow march of trees rarely crosses watersheds this high. A watershed is a precarious stance for a tree, the soil worn thin first by the long, slow, punishing drag of ice, and forever after by the weight of wind and the pull of contrary waters. Perhaps a mischievously blown juniper seed will crash-land on a watershed and take root there, because having landed it does not know how to do anything else or be anywhere else. Such a juniper has the springiness to confound winds. It grows flat and low, and it grows slowly too, for watershed soils are either thin or absent. Black pools in peat hags, rocks that burst through the skin of the land, stag wallows – these are the raw stuff of Highland watersheds, and none is conducive to tree growth. Anything that grows there and that hasn't crash-landed like the juniper is a specialist: dwarf willow (a tree of a kind but about as tall as a rug), sundry saxifrages, cloudberry, bog-cotton and other grasses remarkably adept in the matter of extracting blood from a stone . . . that kind of anything. Any creature that lingers there through the winter has an Arctic inheritance: ptarmigan, snow bunting, mountain hare. Golden eagles and ravens cross watersheds for fun, but very little else does.

So, if you were to walk from beneath the window where

my desk contemplates a view of trees, walk beneath that hill shoulder in the south-west, and follow a furtive road through the trees until you reach a meeting place of glens by an old stone bridge and a waterfall, you would have begun to walk that old way through the hills that ultimately connects Balquhidder Glen to the Trossachs village of Brig o' Turk. Brig o' Turk has nothing to do with Turkey, but is named by way of a collision of old Scots and Gaelic and the historical incompetence of mapmakers, a common enough occurrence where Sir Walter Scott plied his trade, as indeed he did in this of all landscapes. Scott may even have invented the word 'Trossachs' itself, for before Scott's time it was all Menteith, and after his pen had made the place famous it was The Trossachs, and no one knows for sure what the word means although many an amateur Gaelic scholar will give you a theory. But *brig* is just a Scots bridge and *torc* the Gaelic word for a wild boar, and there in the company of the ghost of that great wild manipulator of ancient forest that was the wild boar is a clue to the historic nature of this land.

The way through the bare hills lies west above the birches and alders that thinly and briefly clothe the steep banks of the lower reaches of Allt Fathan Ghlinne. This is a hint of the old order that prevailed before that stultifying Victorian regime finally got to work and almost did for the Great Wood completely. Almost.

The land widens at yet another meeting of glens, a flat green plain that is the centre of a fan of three glens – Fathan Ghlinne, Gleann Dubh and Glen Shoinnie – and where all their waters commingle and swell the girth of the Allt

Fathan Ghlinne, which becomes the Calair Burn at Balli-
more, the River Balvaig at Balquhidder, and the River Leny
for a few miles between Loch Lubnaig and Callander. From
here it swells into the Teith and heads resolutely for the
distant Forth, the Lowlands, Stirling, Edinburgh, and the
North Sea. But you would be walking against the grain of
these remote headwaters of the Forth.

The path turns south and uphill into Glen Shoinnie. You
have already encountered one of the many follies of mapmakers and here is another, for the meeting place is not just
the foregathering of three glens, but four different spellings
of the same word – Ghlinne, Glinne, Gleann and Glen
occur within a couple of square inches of a large-scale map,
yet they all mean the same thing. Shoinnie is another
matter. There is no such word in any of the languages of
Scotland, especially not in Gaelic, which is the language
that named these hills. The nearest thing that makes a kind
of sense is *soinne*, which in some parts of Gaeldom might
have been pronounced soin-yu and in others shoin-yu. But
it means peace, and peace is a strange name to have been
attached to an old Highland glen in this historically bloody
part of the world. Perhaps it was where two warring clans
bartered a peace, or perhaps a clan bard simply thought it
tranquil. The Glen of Tranquillity – perhaps the bard was a
distiller in his spare time. Whatever the why and the
wherefore, the artefacts of the glens' lost civilisations are
everywhere: broken lintels that fell from the doorways of
shielings, sheepfolds, the foot-high imprint of houses in the
bracken, the stones that built them long since comman-
deered and recycled to do the bidding of sheep farmers.

So Glen Shoinnie rises to a watershed, beyond which the landscape slowly changes. You begin to walk downhill, not just into the widening girth of a new glen but also as if into another time, as if every few hundred yards you are greeted by an older century, then another, then another. The reason is trees. Something of the Great Wood flourished here, and then, for the usual reasons it almost died here. Almost. But a widespread and widely scattered variety of native trees somehow survived, and in these turn-of-the-century years since 1996, the Woodland Trust has taken a hand and begun to resurrect something like the woodland that was. In just 15 years the results have already begun to be spectacular.

Glen Finglas is the generic name for a complex weave of three glens: Glen Finglas itself, Glean nam Meann and Glen Casaig, with the broad moorland back of Moine nan Each (the mysteriously named Bog of the Horse) rising to the blunt 2,200-feet mass of Meall Cala dominating everything. Finglas is another one of those names whose origins intrigue and confound scholars; hardly anyone agrees, and your own theory is as likely to be as near the truth as anyone else's. The first thing to say about the name is that it doesn't look right. It doesn't look like a Gaelic word. But then time and scholars – aye, and mapmakers – not to mention the rampant spread of English across the face of a landscape it never understood in the first place . . . all these have long conspired to make a mongrel of the name.

Finglas sounds like it could have originated from two adjectives – *fionn* meaning white and *glas* meaning pale or grey, or green, and it's difficult to think of any reason why coupling two such words would amount to a defining reason to name such a prominent glen. In Scott's time, it was written as Glen Finlas, which, if nothing else, is at least more suggestive of phonetic Gaelic. Take away the 'g' sound from the word and the Gaelic it begins to sound like offers one or two possibilities. The one I like best, bearing in mind the glen has a sunny south-facing aspect and was a royal hunting forest from the mid fourteenth century, is *fionlois*, pronounced 'finlis'. So Glen Finglas or Finlas may (or may not) have been the Glen of the Vineyard.

But above it all, standing on the watershed between Glen Shoinne and the lands of Glen Finglas, I have a choice. The direct route is south down Gleann nam Meann. Or I can go west, contouring across the head of the Horse Bog, then south-west to meet the headwaters of Glen Finglas. However I go, it is the presence of trees in ever-increasing quantities and ever-changing degrees of spectacle that will characterise the journey. The Woodland Trust has put in a 'walking and cycling trail' all the way up to the watershed from Brig o' Turk, and with a huge loop that circumnavigates Meall Cala. If you take on the whole thing, it is 15 miles long and climbs almost to 2,000 feet, a strenuous exploration that begins among some of the densest woodland on the whole estate, then crosses bare hillside above the Glen Finglas Reservoir, back into a more rarefied woodland environment where eerie, scattered survivals of birch, oak, alder and weird hazel groves populate the wide and beautiful Gleann nam Meann, back out onto bare hillside where you slog up over the highest ground, then dizzily back down into the head of Glen Finglas and back into a strange twilight world of hundreds of pollarded alders and birches that stand alone, sometimes a hundred yards from their nearest neighbours. In winter especially, with the hill grasses bleached to the colour of pale straw and the trees bare of foliage, there is something curiously insubstantial about such woodland, as if a ragged army of tree ghosts was straggling away in disorder from an epic battle. I have seen nothing like it. This is not the precursor of a great wood beyond, like the sentinel pines that prefigure the pinewood; this is a whole wood of sentinels.

It is the legacy of long, human habitation. It is strange how much easier it is to read the imprints of our own predecessors when there is evidence of ancient woodland to work with. These pollards' trunks are cut several feet above the ground so that the growth is out of reach of deer, and by repeated and careful cutting they provide a renewable harvest of timber that is almost literally inexhaustible. The oldest trees achieve formidable girths. This kind of ancient wood pasture has a spoor that is thousands of years old, a technique that did not change over countless human generations. Both visually and historically it is an astounding heritage that quite literally has outlived its usefulness, but lives on nevertheless.

In the midst of it all, a grove of hazels suggests that at some point in the glen's human story, the importance of hazelnuts and hazel wood even to the very earliest settlers was formalised into a kind of nut orchard, and it is that which survives here. The ancient past is – again quite literally – at your fingertips. I have always been much more

52

fascinated by the evolution of nature in my native landscape than by the people's story but in Glen Finglas the two are so indivisible that I am carried away by both. We have a rough idea that people were living here 2,700 years ago, but as yet we have no idea what happened before that. It is a long way inland to have been investigated by the earliest of Scotland's post-glacial, seafaring nomads.

Mostly, when I reach the watershed above Glen Shoinnie, my preferred route is down Gleann nam Meann, not least because it has that indefinable aura of all old routes through the hills, routes that were made possible in the first place because the passing of the ice assisted early ideas of communication between the big glens, the broad valleys, the easiest waterways. Gleann nam Meann is, a classic glacial valley where new planting by the Woodland Trust has already begun to hint at how the place will look when trees are restored to every slope and stance and niche where they would once have found their own way. I like to take it at a slow pace, leaving time for diversions up the hill burns among the alders. Nowhere is too wet for alders. They thole not just the spates that overflow the gullies but out on the open hill they face down blizzards, snowfields, landslips and the sheer volume of rain that are the inheritance of so many Highland hillsides. People have known that quality in alder wood for a very long time – two or three thousand years ago it was the timber of choice for the crannog dwellers. They had to anchor their round, timbered houses, so they drove piles cut from tall alders into the bed of a loch, built a platform on those well above the water level, and built their roundhouse onto the platform. Excavations in

Loch Tay have found original piles after at least 2,000 years of immersion, and the spectacular replica crannog near Kenmore uses the same process today.

There is a particular pleasure walking down Gleann nam Meann looking at the scattered pockets of survivors, then having your eye caught by a deluge of new young trees high on the hillside, and knowing that however often you walk here over however many years, the experience will only grow richer.

So that is what lies at the end of the walk from below the window where my desk contemplates a view of trees. But if you were to forgo the watershed approach with its long walk-in and the slow-motion transformation of landscape by degrees from sterility to the slow, uphill march of the trees . . . if you were to decline all that, as most people do, and accept the blatantly visitor-hungry invitation of the Woodland Trust's welcome mat, you would park your car near Brig o' Turk and by the main road from Callander to the Trossachs, and select one of several waymarked walking and cycling options. The moment you begin to walk, hundreds of old oaks and thousands of young birches hit you between the eyes. If Wordsworth had wandered here lonely as a cloud his poem might have been called 'Birches' rather than 'Daffodils', such is the immediate impact of young trees in extraordinary quantities.

The 21st-century Glen Finglas is a reserve, and at almost 10,000 acres, the Woodland Trust's largest. For its scope, its diversity of tree species and tree ages (oaks the size of my hand to dying veterans of two or three centuries, the bizarre, dark geometries of those ancient hazels, crowds and crowds of ankle-high, knee-high, head-high birches, rowans, willows and Scots pines, all of them shepherded by tall, robust survivors many times their age), it is the most optimistic of places to spend time.

All of which heads inexorably towards a significant 'but', and it's this: *but* it is a *reserve*, and by definition a place carefully earmarked and set aside in order to restore its native woodland from terminal decline to rude health; in other words, the rebuilding of a remnant. This is not 'the return of the forest', so much as the return to a remarkable state of health of a living fragment of a forest that almost died. Almost.

If you accept the historic notion of the Great Wood in some shape or form, then you can argue that the Trust is sewing back into place a small patch on the hem of that great filamor of a forest. What attracted the Trust to Glen Finglas in the first place was the survival of a few crucial threads – *old* threads that showed something of the weave of the place when nature was the weaver, something of the diversity of native species and of their preferred stances in that landscape, and crucially, sources of seeds with a pedigree reaching back several millennia to assist the recovery of a native wood. The fact of their survival at all when so many other Highland estates cannot muster enough native trees to make a decent bonfire may have been assisted by its long career as a royal hunting park. Glen Finglas was probably chosen for that role (by David II in 1364) because of the beauty and openness of the woodland, as well as being an easy day's ride from Stirling Castle. Even after the estate was acquired by the earls of Moray in the seventeenth century, the priorities that governed its management remained those of a good hunting park. So when the onslaughts of the nineteenth century began, the Glen Finglas woodland was in a better condition to ward off extinction than many other Highland glens.

Hunters knew from the earliest settlements within the Great Wood that deer needed woodland if they were to survive in good condition. That ancient wisdom was ruthlessly rubbished by the years of the Highland Clearances and the subsequent Victorian invention of the 'deer forest' and the 'Monarch of the Glen' philosophy of deer hunting. And while there is no denying the thrill that a red deer stag on a rocky mountain skyline strikes in a human breast,
it makes for bad biology. Red deer stags in a European mainland forest are a third bigger and carry bigger spreads of antlers. It would be naïve to imagine that the hunting pleasures of the posh people never irked the woodland peasants as they still irk many of us today, especially conservationists and nature writers. In 1707, for example, the earl of Moray's factor unearthed what he suspected was a conspiracy of local farmers to cut down trees and kill deer to create more and better farmland, and offered His Grace the maxim, 'no woods, no deer'. And in the 1980s my friend Don MacCaskill told me 'a forest is not a forest without deer'. And if you could have asked the opinion of thousands of years of Highland wolves they would have agreed too, for it is simply a timeless truth. Deer and trees belong, mutually, and in this of all landscapes it was never truer.

So over the long cohabitation of trees and people and wildlife in Glen Finglas, they have all evolved and prospered and suffered in parallel, if not always in tandem. The nature of the trees dictated the actions of the people, and the actions of the people often modified the nature of the trees. The wildlife tribes made what they could of the changed circumstances; where they adapted they prospered, and where they failed to adapt they vanished. The earliest human settlements depended not just on the availability of wild animals (although that was essential) but also and increasingly on the availability of good grazing for their herds. But the settlers would recognise, too, that the Great Wood already sustained wild grazing herds – notably deer, wild cattle, wild goats, wild horses, the fabulous aurochs – and that all these fostered rich grasslands throughout its entire reach. In the concluding chapter of *A History of the Native Woodlands of Scotland* the authors write: 'It is important for conservationists to recognise that all our ancient semi-natural woods, without any exception, had domestic stock in them on a regular basis. It follows that woods in historic time were likely to have been full of glades which animals kept open by grazing, and where the grass grew best. There are numerous references to fields and even to cultivation in the woods: it is best to think of them as typically mosaics of wood, meadow and bogs.'

None of this is lost on the Woodland Trust in their efforts to regenerate native trees throughout their entire natural range in Glen Finglas, a range that extends from the edge of the ominously named Black Water marshes west of Loch Venachar and the Brig o' Turk mires, to the blasted acres of Meall Cala and Moine nan Each almost 2,000 feet higher. A working farm maintains the practice of grazing, although the numbers of cattle, sheep and red deer are now carefully controlled. The *raison d'être*, after all, is to grow trees. Deer and trees will find a workable balance in time, even if ultimately it takes the reintroduction of wolves to demonstrate nature's true efficiency in that precarious art. But as an object lesson for the immediate wolfless future I well remember the Nature Conservancy Council's ground-breaking decision to buy Creag Meagaidh above Loch Laggan in Inverness-shire, back in the 1970s. The plan was to spare it the terrible fate planned by its owners, Fountain Forestry, who had recently bought it, and to nurture its entire range of montane vegetation from lochside to plateau, and in particular an outstanding mountain birchwood. Shortly after the takeover I interviewed the NCC's Dick Balharry who masterminded the work on the ground. Given the widespread criticism of what many saw as a heavy-handed red deer cull, I asked him how many deer the NCC was prepared to cull to achieve its ambitions. His response was that they weren't counting deer, they were counting trees, and that when the deer population could be sustained by a vigorous and healthy birchwood, that was how many deer they would cull. You see, people and woods have interacted forever, and they still do.

So venison, lambs, calves and – of course – timber are products of this latest evolution of human settlement in Glen Finglas, and that too honours the oldest woodland traditions of our species. There are other restoration and new planting projects around the country although none is on this scale. The Glen Finglas restoration is praiseworthy work, but it is local. Its influence does not extend beyond its own boundaries, does not cross watersheds. At least, not yet. The Great Wood of Caledon, if it ever lived up to anything like the historical hype, must have been a Highlands-wide forest that characterised at least the heart of the mainland Highlands as much as did the mountains. Even if the legend of its past is only half true, those surviving woods that we presume to be its remnants are too far-flung and too far apart to be environmentally meaningful today as any kind of entity. The Great Wood of Caledon has the same fabled ring to it as *Tír nan Óg* or the Loch Ness Monster. But watching Glen Finglas over four seasons and several years has put the possibilities of belief within my grasp . . . the eagerness with which this rugged land responds to trees, and to the slightest encouragement from the human race . . .

Whenever I walked in Glen Finglas, watching and being intrigued by what unfolds there, being flabbergasted at just how eagerly the trees respond and transform, I kept thinking: if only we could safeguard and enlarge the woodland strongholds, those surviving enclaves of trees that bridge the chasm of the millennia by their very survival in the very landscape where their first elephant-slow march across the land set down their earliest ancestors. And then, if only we could connect them up with patches of new forests. And why can't we enlarge the strongholds and connect them with linking groves of new trees? Why can't we put it back? If all of those bare glens beyond the

watershed to the north of Glen Finglas could be thought-
fully wooded again, mimicking nature's old design – linking
Glen Finglas to Loch Voil's wooded south shore in Balquhidder Glen and the Strathyre Forest along both shores
of Loch Lubnaig, where there are also significant oak
woodlands among the Commission plantations – something
that amounted to rather more than the sum of the parts
would begin to take shape. Then the Trust acquired a piece
of land that might achieve something similar but at the
southern edge of its estate, extending it right along the
north shore of Loch Venachar to meet the southern end of
the Strathyre Forest at the foot of Ben Ledi.

And then, it seemed to me that the very location of Glen
Finglas assumed a critical importance. I now think that if
there is one place in the southern Highlands that can
become a showpiece not just for demonstrating the worth
of an expansive tract of diverse native trees for its own sake,
but also for making the argument for a born-again Great
Wood that blends enclaves of native trees with much higher
standards of plantation forestry . . . that place is Glen
Finglas. Loch Achray and Loch Venachar lie to the south
of the Woodland Trust land. West of the lochs lie the
Trossachs oakwoods and Loch Katrine, while south of
them is the Queen Elizabeth Forest Park, which is the
Forestry Commission's showpiece at the heart of the Loch
Lomond and the Trossachs National Park. At their best,
these Commission woodlands blend old and new pine and
oak and birch woods with spruce and larch plantations. But
nowhere in the huge scope of the Queen Elizabeth Forest
Park is there anything on the scale of Glen Finglas to reveal

the purity of purpose of ruthlessly native woodland restoration. It is the company Glen Finglas keeps that will
permit the shining example of its fully-fledged woodland to
soar beyond what even its most ardent champions envi-
saged – that and the fact that it drives the impetus of the
forest world northwards into wilder hills with no main
roads, and to within a few miles of Balquhidder. If I were a
rich man (instead of a nature writer) I would happily put a
lot of money on a future which demonstrates that when
Glen Finglas is finally replete with native trees of all ages
growing throughout their natural range on the estate, it will
prove infinitely richer in every imaginable form of wildlife
and infinitely more satisfying to people as an environment
in which to spend time and marvel than anything the
Forestry Commission has to offer.

The value of Glen Finglas is that it shows what is
possible. The sum of what is possible Highlands-wide,
and Scotland-wide for that matter, is limited only by our
imagination, by our daring, by our vision and by the extent
of our willingness to realise it, and our willingness to pay for
it. If you are as helplessly in love with the Highland
landscape as I am, and not a Roman legionary pushed
far beyond the extremity of your comfort zone, you may
judge the results of such a vision to be one of the soundest
investments that 21st-century Scotland will ever make.

CHAPTER FIVE

Sunart

'The woods, of which the oakwoods are the dearest and Highland-rarest, are a deceiving, coast-hugging, burn-clinging throwback, an echo of the landscape that was. The richness of the meagre thrivings only underpins the sadness of the few moor-stragglers about their fringes, isolated trees set eerily against a sea-rooted ground mist in autumn dawns. It is winter for the oakwoods, however, for these are the last of their tribe.'

I wrote that in 1989 in a chapter about Ardnamurchan in a book called *West Highland Landscape*. I now cheerfully eat my words; a second spring for these very oakwoods has begun.

In the same chapter I also wrote this:

The shores of Loch Linnhe's narrows are chalk and fragrant cheese, the east burdened with the loud and sprawling implications of Fort William and the trade-

and-tourist route north; the west a sparsely tramped, sea-glittered, oak-dusted way which dawdles and dwines to the edge of the sunset . . . So for those thirled to Ardnamurchan, whether by birth or by its luring land-scape, the Corran ferry's crabwise gait is imbued with heavy significance. Alasdair Maclean, poet of Ardna-murchan, whose *Night Falls on Ardnamurchan* is a well-crafted milestone in the literature of the West, calls it 'a kind of mobile decompression chamber where various kinds of pollution were drained from the blood and I was fitted to breathe pure air again'.

It is a besotted native's perspective as well as a poet's, rather than, say, a dispassionate scientist's conclusion about the relative air qualities east and west of the Corran Narrows of Loch Linnhe. It's a crossing that does not quite last five minutes, but there is no denying the instant change in the landscape, the culture, the very feel of the journey, and yes, the frame of mind of the traveller who heads west from the ferry. And the converse is also true. Crossing back east to the rumble of the A82, the unpre-possessing squat of Fort William at the base of massive, cloud-shrouded mountains, or the dreaded psychological impact of the main road south (any main road south any-where in the Highlands), is, I suppose, and using the poet's yardstick, a compression chamber where travellers re-inhale all that is less desirable about the land beyond.

But west of the narrows can be a confusing land to strangers. Signs in both Gaelic and English welcome you to both Morvern and Ardnamurchan, and if you don't quite understand how you can be in both at the same time, well that's just the way it is, and you're probably also in unsignposted Sunart and Ardgour. Ardgour is the one you leave first for it lies to the north of the village of the same name where the ferry lands. As you go west, Sunart is where you start to notice oak trees. The long arm of the sea called Loch Sunart bisects the three-in-one land of Ardnamurchan, Sunart and Morvern, and it does at least drive a wedge of clear distinction between Sunart on its north shore and Morvern on its south, but it is far from clear where Sunart ends and Ardnamurchan begins. Except that Alasdair Maclean and many other natives will have you know that it's all Ardnamurchan, and that Ardnamurchan Point with its famous lighthouse 50 miles west of the Corran Ferry is only famous because it marks the westernmost point of the Scottish mainland, whereas it is really just one more rock among Ardnamurchan's millions of rocks.

The only other thing to bear in mind, and it will do nothing at all to ease the confusion, is that a programme of native oakwood restoration and management and enlightened community endeavour that extends from Moidart in the north (north of Ardnamurchan and west of Ardgour) to the Sound of Mull in the south and to Ardnamurchan Point, is known collectively as the Sunart Oakwoods. Whatever you call it, and whatever the names of its constituent parts with their infinitely flexible boundaries, it is as beautiful and rarefied a land as any in the Highlands, which, at its own leisurely oaken pace, is growing woody and green again, and making something of a name for itself as a partnership between nature and people. And yes, the more you familiarise yourself with the place, the more you inhale its aura and its good air, the more you comprehend utterly where the poet was coming from. Between the village of Ardgour and Ardnamurchan Point the unfolding tapestry of landscapes has wrought more raw emotion from me than any other comparable distance of mainland Highland miles anywhere. One early autumn I watched the tail end of a hurricane – a proper hurricane with a name – take a swipe at the lighthouse and the rocks of Briaghlann. I saw the sky fall on the ocean black as night at midday, saw the sea drain of all colour but grey and white, saw the shape of the wind slice lumps from endless queues of waves, saw the lighthouse blur as the smashed sea leaped for the throat of the banshee. And in the midst of all that two cormorants sat riding out the storm; no point in flying in this, so sit on it. Pipits, curlews and other waders hid on a leeward ledge, and two whooper swans newly arrived from Iceland hugged the shoreline of Loch Grigadale, visibly unsettled by the fury of everything. It was a long trek down the northern ocean for such a landfall.

In half a day the thing was done and Ardnamurchan re-emerged as the rain stuttered and fragmented into soft showers and a warm wind blew through evening rainbows. Every cottage storm-porch reeked of dripping oilskins. Ben Hiant ran white with temporary waterfalls that were gone by dawn. The morning after, the first passengers off the ferry found the place enchanted and wondered at the fuss the natives make of the weather.

Being a late twentieth-century poet, Alasdair Maclean was from a generation accustomed to entering and leaving this land by a car ferry that connects one part of the mainland to another. Even though the Ardnamurchan peninsula west of that neck of land between Salen and Kentra feels more like an island with a land bridge, the place presents such an out-on-a-limb aspect to the rest of the world because it takes such a long overland journey from almost anywhere to get there. It was not always thus.

The long Atlantic inroad of Loch Sunart made this one of the most accessible of lands among those countless eras of seafarers who variously explored, exploited, settled and sold out the Highlands' western seaboard, and it did so for all those eras until the most recent chapters of Scotland's long history. Travelling overland to Ardnamurchan was an absurd proposition to them all, none more so than those who may have summoned to their homesick imagination the notion of the Great Wood. And here was a kind of ceremonial gateway to such a place – miles and miles of shoreline where the trees descended to the shore in waves and displayed themselves there like defiant armies, wading out among rocks into the shallows at high tide, or here and there standing as densely packed as spears in a schiltron around the landward edge of a sandy beach, and even clothing some of the sea cliffs. If there was a single frontier of the Great Wood calculated to intimidate the weary, the unwary or the susceptible minds of thousands of years of seafarers, perhaps it was this one.

The oakwoods still wade into the Atlantic today but it is a toe-in-the-water frontier of the Great Wood now, for as with all great expanses of our ancient native woodland, most of it has gone. But it is our very own era that has resolved to give the oakwoods a future, to put them back at the heart of community life, to forge a new pact with nature, to give back where so much has been taken from nature for so, so long, to restore as much old oakwood cover as is humanly possible, to raise from the last gasp of its own seedbed a remnant beyond price of the temperate rain forest of many tree species that once grew all along the western Atlantic coasts of Europe. To see its like now you should go to the south-east panhandle of Alaska where you can walk for days in trees, then come on a huge, dead-still lake, only to be started by the sudden emergence 50 yards offshore of a humpback whale – the 'lake' is an arm of the ocean, a Loch Sunart of the Pacific.

The potential of Sunart was identified by a system of Special Areas of Conservation established under the European Union's Habitats Directive. Local and national Scottish agencies formed the Sunart Oakwoods Initiative at the beginning of the new century and the Great Wood breathed again. The local community has devised ways of managing the woods that are both generous to nature and still permit the use of timber for houses, fuel, fencing, crafts and boat building. The local tourism industry is emphatically green, and there is nothing quite like it the length and breadth of the Great Wood. But then there is nothing in the Great Wood quite like the Sunart Oakwoods either.

Ariundle, just north of the likeable lochside village of Strontian, is the Shieling of the White Meadow, apparently a strange name for an oakwood, but then there have always been meadows in the Great Wood, and from the times of the very earliest woodland settlements there were areas of cultivation. More prosaic is the Norse Egadale – the Oak Glen. Interesting: the alien invaders saw a land of oak, while the native Gaels distinguished between the local character of individual shielings.

I paused here to walk knee-deep in my own memories, for I have a personal history with this land west of the Corran Narrows. I first got to know the name of Ariundle as a sign that my journey to a particular destination was almost over, except for the worst part of it, which was the single-track road beyond Ariundle over a wicked hill pass and down the far side to Polloch whence it degenerated into a forest track along the shore of Loch Shiel. The purpose of the journey, which I made three or four times, was to meet wildlife writer Mike Tomkies who would boat across the loch from the home he called Wildernesse on the roadless, trackless far shore. My great friend – which he was in those days – was generous, hospitable, amusing and instructive company whose work on golden eagles in particular was – and I believe still is – unequalled in Highland Scotland. The best of it all was the time I helped him with his work at an eagle eyrie at the far end of what he called 'the Killer Trek', which I have written about before,* and which I still often recall when I walk the eagle hills of Balquhidder. Much of what I put into practice now on these home hills I learned from listening to or just watching Mike Tomkies at work. And over the years I saw him regularly my admiration for golden eagles became a love of golden eagles, and that was his lasting gift to me.

* See *Among Mountains* (Mainstream, 1993)

We were sitting on a hillside on that trek having a bite and a breather when he told me that he didn't care what 'they' said, he was utterly convinced that mountains like these (he indicated with a broad sweep of his arm the bare peaks to the north of us) were never wooded, that there never was a 'Great Wood of Caledon' in such places. It was the first time I had heard the old orthodoxy challenged, and challenged at that by someone whose relationship with wild country was more *instinctively* true than that of anyone else I ever met, as if he possessed some ancient landscape wisdom. Perhaps it had come down to him through his Islay mother.

Until that moment, I had never thought that much about the Great Wood of Caledon. It was a phrase that people of a certain age used in a certain way that implied a halcyon time of forest Nirvana. Frank Fraser Darling, once widely regarded as the greatest of twentieth-century naturalists and a founder of the Nature Conservancy Council, even wrote in 1947 that 'the imagination of a naturalist can conjure up a picture of what the great forest was like: the present writer is inclined to look upon it as his idea of heaven'. As late as 1991, the great wildlife film-maker Hugh Miles made a TV documentary on pinewoods and collaborated with *Sunday Times* journalist Brian Jackman on a companion book which exclaimed that 'there was hardly a glen that was not roofed with trees, the high hills rising like islands from the blue-green canopy', and here at around the same moment was Mike Tomkies denouncing the idea. So, more or less from that day forward, I went in search of my own ideas about what the Great Wood may or may not have been. Historians have been disagreeing about it for 150 years since it was branded 'a myth' by Cosmo Innes, a noted Victorian authority. I had decided soon after my chat with Mike Tomkies in the mountains above Loch Shiel that I should have an opinion but that I would take my time forming it. Wherever I travelled in my constant and characteristically restless explorations of my native land I would look out for what survived that was demonstrably old. I would try and read the land as well as those history books I trusted, and see what rubbed off. I wanted an opinion because I was sure that sooner or later the question would come up; twenty years later, it did, and I was asked to write this book.

*

So I had crossed the Corran Narrows and headed west, turning aside at Strontian for Ariundle to walk in its oakwood and to renew old acquaintance. The thing about mature oak trees – or one of the things about mature oak trees – is that they insist on their own space. Size does not always matter in my appreciation of trees (I love aspen, birch and rowan, for example), but sometimes I like to be reminded of the sheer presence of a truly colossal tree, not tall so much as massive, with limbs like tree trunks and a tree trunk the girth of a thicket. I remembered such a tree in the Ariundle wood but I had no memory of how to find it.

It was the first day of spring and the oaks were bare and fat-budded. Limbs, branches, twigs, twiglets and buds were deep black against a pale and tattered blue, white and grey sky that fired icy squalls down on the morning with short, deceptively sunny calms in between. Looking up at a big hillside oak (big, but not massive, not *that* hillside oak) from below the level of its roots was like looking up at a stained-glass window, albeit one whose artist had worked with a limited palette – just these three shades, all of them pale, none bold, none vivid. The teeming tracery of the twigs was the leaded tracery of the window. If stained-glass windows were like this, I thought, I could be persuaded to go to church. Nature writers have often reached for the analogy of a cathedral over the years when they tried to write down a forest, with the tallest, straightest trunks as the pillars. I think the notion is back to front; the cathedral is like the forest. Surely a stonemason with big ideas in his head but lacking the means of their execution walked into an ancient wood one day a thousand years ago, looked at a pair of tall, straight trunks and thought, 'Hmm, pillars'. And another (or the same one on a different day) looked up at a big oak on the first day of spring and saw the sky through its thousand branches and thought, 'Hmm, stained glass'.

Any old wood that accommodates big oaks will also be an open wood, because the nature of oak trees insists on it. Besides, people have been working these woods to some extent or another for 8,000 years, and 'working' means felling trees, creating spaces (for settlement, for cultivation, for charcoal-burning that was an essential part of iron-smelting), and think of it – an oakwood on the edge of a long and sheltered arm of the sea – what a gift for repairing and building boats.

Nature, of course, had been working the woods for rather longer, and changes wrought by climate, fire, storm and flood all manipulated a forest like this, discouraging some species, washing up new ones, readjusting the ground cover, the insects, the birds, the mammals. The oakwood I walked in on the first day of spring is a much-tampered-with piece of ground, for all its aura of timelessness, and the tampering goes on to this day, although ours is arguably the first of all this wood's eras to approach it in a mindset of healing.

The only oak leaves not on the ground were a few bleached and wrinkled souvenirs of last year that, for reasons I don't begin to understand, clung to their parent tree through a long, turbulent and particularly cold winter. It is a feature of oak trees, and especially young ones, that when the new spring's leaves begin to unfurl, they are briefly outnumbered by last year's tenacious dead. Yet even on 21 March the big trees are green – the north-facing curves of the trunks and the great limbs are simply plastered with moss. Moss grows here in a swaddling electric-green fur. Many of the biggest oaks at Ariundle have rooted among rocks. The moss has a particular liking for them, has coated them with such a depth of green fur – six, seven, eight inches – that they all have the same smoothed-over shape, and these rocks look more like broken pieces of tree than broken pieces of mountainside.

Sometimes oaks rooted in old drystane dykes. The growing strength of the tree first prised the stones apart, bursting open the dyke, then simply grew over it on both sides. The dyke fell into disrepair or perhaps its stones were recycled to build a house or a barn, and the tree simply grew and grew among the few scattered traces of its birthplace until the moss finally moved in and united them in a common oakwood uniform of electric green.

There is a second shade of green about a big old oak on the first day of spring, or a midwinter oak for that matter – ferns. They grow in and around the forks, and out along the big limbs in loose clusters, and they wave in the wind and soften the huge, eerie, bare blackness of a living oak seen against the sky, for all the world as if they were the foliage of the tree itself instead of – well, parasites, albeit benevolent parasites.

I found a tree that offered a comfortable backrest and sat. I like to write where my raw material is raw. The best place to be when you write down an oakwood is in an oakwood. It is hardly a new idea. Hereabouts, there was an eighteenth-century Gaelic poet, a contemporary of Burns, called Alasdair MacMhaigstir Alasdair, otherwise known as Alexander MacDonald, though I much prefer the unadulterated original. For obvious reasons, one of his woodland poems graces the Sunart Oakwoods Initiative literature. It is called '*Oran an t-Samhraidh*', 'Song of Summer'. It celebrates the first day of May, and is reproduced both in the Gaelic original and an anonymous English translation I didn't much care for. Anyway, I had scribbled it down earlier and now I pulled out the notebook in question and frowned at the offending page from my throne at the foot of the oak which prodded me between the ribs and said, 'Go on then, you do better'.

Song of Summer
(after Alasdair MacMhaigstir Alasdair)

The feadan* of the woods
woke me early, the sun
brightened even this shady hollow;
the rocks warmed to sweet vapour
the dew, echoed
with an elegant rebound
the Piper's threnody;
the dew-lit buds echoed
the dance of a thousand suns.

So often, when you walk alone, deep into a wild land-scape, and then find a quiet place to sit, and if you have the gift of sitting quietly and still for a while, and especially if you are wearing clothes that are something like the colours of that landscape . . . so often the natives of that place treat you like a piece of the landscape. I had been still for a while wrestling with the mindset of the eighteenth-century poet and enjoying the company of the trees, when I lifted my head to follow an instinct, the kind of instinct which is born from innumerable hours over many years in just such a situation as that one. I had followed a skinny and inter-mittent path before I turned aside to sit at the tree, a tree

---

* *Feadan* – the chanter of the Highland bagpipes. I was unclear if his use of the word in the original meant literally the playing of a chanter, or if he was using the word to indicate nature's music on such a morning, so being a poet I chose the latter, because being a poet he did too!

about the same girth as a roe deer. The instinct that now demanded all my attention was that something had just changed. I had yet to register whether sound or movement or a stray hint of wind-blown scent had effected the change, but it had nudged me away from the *feadan* dilemma and demanded that I be alive to every nuance of the moment. One of the virtues of sitting still in front of a restricted view is that you very quickly become intimate enough with its broad brushstrokes and many of its details to notice when something is not exactly as it was the last time you looked. So, without moving, I worked my way through the close quarters of the wood, then the middle distance, then the furthest extent of what was visible, which at that moment covered less than a hundred square yards. Nothing.

I tried again. The only obvious movement was an orange-tip butterfly crossing a patch of sunlight, and I watched it for a few yards of its wandering flight about a yard off the ground. But I knew that was not the source of the change. The butterfly dipped towards the path then suddenly and apparently decisively changed course away from it, and as it did so I caught a new movement, dark and low down. I focussed binoculars on the spot, and all that showed that was not there before was what looked like a couple of mushrooms stuck to a fragment of dark wood that caught the sunlight, except that you don't get mushrooms on the first day of spring and there was a glossy hint to the dark wood. Then the dark wood moved and rose a few inches, the mushrooms materialised into ears and the dark wood into the deep brown face and black-eyed stare of a pine marten at 20 paces. The wind was briefly in my favour, but in woodland like Ariundle it can bend anywhere as it seeks a passage through different densities of trees. The marten moved forward with the stealth of a hunting cat. As he moved he climbed a short slope on the path, and his peach-coloured throat and creamy shoulders and upper forelegs caught the sun and made him splendid. I think he knew that, and I think his new stance was calculated to impress. Because I have seen that stance before.

The building at Balquhidder, where I have had my writer's eyrie for a few years now, has a back door that opens out onto a long back garden, mostly grass, with a birch and spruce wood beyond. I was standing there one day deep in conversation with a friend when a pine marten appeared at the far end of the garden and advanced on springy legs to within four or five yards of where we stood. At that point it decided to pay attention to us, and it did so by fixing us with small black eyes, advancing a yard out of shadow into sunlight and (I can think of no other image to describe his action) *flourishing* himself for our benefit. He was making a show, and he was as unintimidated by two men as any wild animal I have ever seen. Having made his point (whatever it may have been), he dived under a hedge beyond which lay his probable destination – the nearest rubbish bins.

So now, with my back to a venerable oak in Ariundle on the first day of spring, there was a swaggering familiarity about the blatantly struck pose of one more pine marten. I realised I was grinning, and I felt as if I should applaud. Instead, I said out loud,

'Hello stranger.'

Then the orange-tip was back, and it flew down to within a few inches of the pine marten's nose, and the animal's eyes followed that instead, but it danced upward and away, and vanished among the trees, and the pine marten gave me one more unconcerned glance and ambled quietly in the direction of the butterfly. And there was one of the oldest encounters of the Great Wood – the man and the pine marten; we have been meeting each other in circumstances just like these for thousands of years.

It is not so long ago that the pine marten was holed up in a desperate battle with extinction on the Sutherland coast and nowhere else in the land. It is not so long ago, either, that the Sunart Oakwoods were similarly pinned down and apparently doomed. But our species pulled out of both offensives. The pine marten was given a surreptitious and wholly unauthorised helping hand to colonise other Highland woods, an invitation it has responded to with great eagerness. Its recovery has been astonishing.

It can also become very tame. Once, at Mike Tomkies's cottage, I sat a yard away while he coaxed one into the room through an open window and watched with disbelieving eyes as it gently took a piece of bread and strawberry jam out of his mouth. I cannot say I enjoyed the spectacle, but it taught me something fundamental about this extraordinary creature: it is utterly fearless. It also punches far above its weight. My car headlights once illuminated one on a quiet back road in the Trossachs. It was trying to drag the carcase of a young roe deer from the road into the undergrowth, and very slowly, a few inches at a time, the carcase was moving. When I returned a few hours later, there were only blood-stains.

As these oakwoods rejuvenate and expand, the scope begins to multiply year after year for pine martens and much else besides – including orange-tip and the much rarer chequered skipper butterflies. People are learning something else in the benevolent climate of such thoughtful restorations. It is that creating new opportunities for nature – or restoring old ones – creates new opportunities for people too. Expanding oakwoods for nature does not mean that we cannot use them too. The people who live and work within the Sunart Oakwoods are learning that at first hand. I was reading about some of the initiatives while I enjoyed good coffee and a bacon roll in the craft centre at Ariundle, and the spirit that it all evoked reminded me of Alistair Scott, once a senior manager in the Forestry Commission, and the author of a likeable little book called *A Pleasure in Scottish Trees*. Its introduction, which he described as 'my credo about trees in Scotland', concluded:

Finally, I like wood. Wooden houses, wooden barns, oak roofs, pine telegraph poles, cedar shingles, clinker-built boats, picnic tables, chairs, axe-handles, chopping boards, bowls, spoons, spurtles, totem poles, Morris Travellers, Russian dolls. I delight in the way that every culture has put trees to appropriate uses. It is a universal pleasure. How satisfactory, if we could reconnect the trees growing out there with the wood that pleases us in here. And start to innovate for our own twenty-first-century needs. Hence my admiration for the Wood School at Ancrum, Jedburgh – half a dozen young designers and craftsmen in wood, sharing the expensive machinery. If we are going to be better Europeans, could we not be Scandinavian in our use and appreciation of wood?

I found the huge tree I was looking for and remembered it anew. It seemed to have grown to a new dimension – of immensity – since I had last seen it, about three years before. I won't insult it by guessing its size or measuring its girth, but Alistair Scott mentioned totem poles, and this is how I like mine. The totem poles of Native Americans held rituals of the tribes that made them. The tribes of Highland Scotland – the clans – drew many of their rituals and traditions from nature: everything from the letters of the alphabet (named after trees), the clan crest (a flower or a tree leaf), to the golden eagle feathers that adorned a clan chief's bonnet. The oak, being the most massive native tree in the Highland landscape, the wood of choice for roof beams of the great castles, and therefore symbolic of supreme strength, was a natural emblem for one of the strongest clans, Clan Donald. But there is more to it than that. If you were to ask me to symbolise the Great Wood, I would bring you here. I would stand you before this oak tree in this wood of oak trees and ask you to marvel. For this is what nature is capable of when we allow it room to breathe – this workaday miracle, this routine giant. It looks exceptional to us because all across the Highlands, all across the landscape of the Great Wood, we have denied nature the opportunity to make such trees. The Sunart Oakwoods project will surely be a new beginning that redresses that grotesque imbalance.

*

Kinlochmoidart is one of those names that makes a kind of music on your tongue. There are Kinlochs all over the Gaelic-named landscape. The 'Kin' bit is a crudely angli-cised *ceann* meaning 'head', and almost every Kinloch you ever saw stands at the head of a loch, in this case Loch Moidart. There is another Kinloch at the head of Loch Teacus that opens into Loch Sunart, another on Loch Morar, another on Loch Hourn, and so on all the way up to the Kyle of Tongue on the north coast of Sutherland facing Orkney (for the Kyle is a sea loch despite its name, and Kinloch at its head proves it). So Kinlochmoidart is also one of those names that does what it says on the tin as well as making a kind of music on your tongue. I had a spell in what now feels like an earlier life when I used to wash up at Glenuig a few miles up the road, where I had found a perfect, sheltered ledge a few feet above the high tide with room to pitch a tent, where I grew accustomed to falling asleep and waking up to the sound of Arctic terns. Like Ariundle, Kinlochmoidart was a name on a signpost that meant the familiar journey was almost over. Now, it too means oakwoods.

The transformation in my awareness of the place came about thanks to a week in a small, old and isolated cottage in a grassy clearing in the middle of an oakwood. The atmosphere of the place was unlike anything I had ever known. There is a self-containment in such places that discourages further travel once you arrive. Something like it must have impressed itself on the earliest settlers and encouraged the very act of settlement. It is an introverting place; the trees shut out the mountain landscape beyond, and, although you could walk to the shore of Loch Moidart in less than half an hour, it is a narrow stretch of water, enclosed by the bulk of Eilean Shona, and the oaks wander down below the highest tide line.

If you just sat around outside the cottage, sooner or later all the wildlife of the wood would come to visit. A pine marten jumped onto the kitchen windowsill. An entry in the cottage visitor book exhorted me to 'feed Bobbie' with bread and jam, and while I felt inclined to add a footnote to the effect that its name is pine marten, not Bobbie, and it does no tribe of nature any favours to be treated like a pet, I put food out for him and watched quietly from the kitchen while he ate a yard away on the other side of the glass. Roe and red deer were daily visitors to the clearing, especially in the early morning. The list of bird species made impressive reading but the one that surprised me turned up on the second afternoon while I sat outside with a coffee and a notebook. A thrush had just crossed the clearing calling loudly and flying with unusual haste so I gave it a second glance, and was just in time to see a buzzard-sized bird swerve between two trees at the edge of the clearing and follow the thrush back into the trees with a similarly deft swerve not six feet off the ground. It was the behaviour not of a buzzard but a goshawk. I saw it almost daily after that, sometimes in the edge of the clearing, mostly slipping through the trees with the agility of a bird a tenth of its size; one more workaday miracle for the watcher in the oakwoods.

I learned something else in the course of that memorable week, and it had to do with a recognition of Don Mac-Caskill's observation about 'the people of this country, in whose history there is so little of a forest background'. It is so true; so few of us live in the woods, as deeply immersed in the woods as that cottage, so that every hour you spend there is an hour in thrall to trees and all nature that keeps them company. That is what we lost over the last 5,000 withering years. I had the briefest glimpse, the most tantalising taste of how many of the earliest of Highlanders must have responded to their surroundings, although of course at any moment I chose I could get in the car and drive beyond the reach of the trees. Yet I felt disinclined to do that, and for quite a while after that week was over I missed the company of the woods and the sense of living in a clearing that someone had made long ago so that they too could keep that kind of company.

I was back in that same wood a few months later in the snow, en route to Mallaig and Skye. I didn't stay in the same cottage but I breathed in the same atmosphere, and there were foxes barking at night and that deep, deep contentment, that suddenly remembered self-containment, settled on my shoulders again.

Quiet Tonight

Quiet tonight, not silent
– sideways snow
made hasty whispers
at the window, not still
– rummaging winds
ballooned the flakes
and rattled the black oaks.
What made it quiet
was the pallid darkness
of the snow-stuffed sky
and the shy lull
that followed fox bark
after fox bark
after fox bark.

CHAPTER SIX

Strath Fillan

The track looks old and well made. It is stony, but with a strip of grass up the centre and along each edge. Think carts, towed by horses, think clansmen in sandalled feet. Oh, go on then, think stalkers' Land Rovers if you must. I like how the track rises and falls and leans in and out over the small contours, how it has been hand-made to fit the land. It skirts a primitive tract of bog and glacial moraines. The River Fillan is what remains of the glacier in question; Strath Fillan is its spoor, defines the breadth of its passage.

A small and furtive watersheet lies amid the glacier's heaped footprints. It is called Lochan nan Arm. If you happen not to be blinking and if you happen to be looking at the precise moment in precisely the right direction, you might catch a glint of gunmetal grey or (much less likely in this rainiest of airts) of reflected sky blue, and puzzle over what might lie there. In which case you would not be the first, for its name means the Lochan of the Weapon and it is the last resting place of a sword belonging to Robert the Bruce who came off second best in a skirmish here and ditched the sword to facilitate a hasty retreat. Or so they say, dismissive of the obvious riposte that such a fighting king throwing away his sword sounds a touch unlikely, a touch like surrender, and the Bruce did not get where he is today with a taste for surrender in the face of adversity. Anyway it was all a few years before he became the master strategist who turned a little-known lowland stream called the Bannock Burn into a torrent at the heart of the psyche of every Scot from that day to this, and finally united his fractious kingdom.

It may be that he always intended coming back to retrieve the sword, but there is no evidence he ever did. On the other hand, there is no evidence he ever threw it in there in the first place, not that that has stopped many a treasure hunter from dredging the place from time to time. In any case, I imagine that the Bruce had more than one sword, being king and all that. And anyway, he was more of a battleaxe man, wasn't he?

The track has been jacked up abruptly to cross the railway, beyond which it dips again then curves, then breasts a rise. You then stop dead at the sudden, deep green blessing of an old pinewood, one more souvenir of the march of time since that long lost day when Strath Fillan's glacier gave up the ghost. One way or another, great age comes at you in waves hereabouts.

The wood is one of the larger living souvenirs of the post-glacial period, perhaps a square mile, large at least by the standards of most remnants of the Great Wood in the southern Highlands. It catches your eye at once and commands it to linger simply because it is such a rarity in the landscape you travel through, simply because it is a landform that belongs, simply because it is a glimpse of an older order, a much, much older order. Yet nowhere within are the trees darkly dense enough to blot out the surrounding mountains the way that a wood like Rothiemurchus (which is ten times its size) can mask the entire mass of the Cairngorms from time to time. The thing is, both these pinewoods are relics. Five thousand years ago, give or take a thousand either way, Strath Fillan's pinewood probably sprawled all the way to Glen Orchy, rolling round the ends of the hills, and Glen Orchy was the centre of a more or less continuous forest that extended from Glen Etive in the west to Rannoch in the east, clambering over and round the lower slopes of many a mountain; Rothiemurchus would have been a component part of a greater forest that swaddled the lower slopes of the whole Cairngorms massif. Now it's the other way round, and the mountains swaddle the isolated relics.

The Gaelic name for the Strath Fillan pinewood is Coille Coire Chuilc, which would seem to mean the Wood of the Reedy Corrie, but there is not a reed in sight. It is, of course, quite possible that there were reeds in abundance when the place was named, however many thousands of years ago that may have been. The name, like the old forest core, endures, but the details of the landscape – any landscape – come and go as it ages and warms and cools, and grows wetter or drier. Maybe the reeds disappeared with the last of the beavers, for nothing manages and expands and sustains wetland like beavers, and they have been gone from here for two or three or four or five hundred years; as with the last wolf, nobody really knows.

As you close in on the forest you begin to pass its outliers, solitary Scots pines designed by the wind (and one of them redesigned by lightning, for up there a hundred yards above the track is the very wounded pine I identified in the Prologue). These sentinel trees are conspicuously short with wide trunks, and I find them curiously affecting. I admire their stoic stance, their very defiance. These are what Seton Gordon had in mind when he wrote: 'I do not recall ever having seen a forest outpost uprooted by the wind . . . they stand undismayed against gusts which send their fellows farther in the forest crashing to the ground.' And it is true, you never see them felled, and you remember again Muir's claim of immortality for the Sierra juniper or Peattie's hint at the same quality in the giant sequoia, 'only a bolt from heaven can end its centuries of life'. And sure enough, there is the tree that I have known for nearer 40 years than 30, one that I thought the handsomest in the whole of Strath Fillan until 'a bolt from heaven' split its northernmost limb from the trunk, so that over the ensuing years the bark fell from the broken timber and it turned ashen-gray and died. Yet the tree itself declined to die, and there it still stands disfigured by its useless gray claw, but its roots still dig in for immortality, and in all the landscapes of my life it is among the most precious of landmarks.

The track dips towards the rush of a mountain burn that flanks the pinewood on its eastern edge, and which must account at least in part for the remarkable survival of the wood because it is difficult to ford at the best of times and impossible in spate. There are two flimsy, greasy and gap-toothed wooden plank bridges a few hundred yards apart, neither of them for the faint-hearted, and over the years I have found drowned sheep and deer in the burn. I imagine many come to the bank and simply don't try to cross. The track curves away from the wood before it reaches the burn and begins to climb. Here it loses the patina of age, for at some point in its not so distant past it metamorphosed into a forestry road to service the modern spruce and larch plantation above the pinewood, a plantation that thought-lessly precludes the possibility of the pinewood ever spread-ing uphill.

As the track climbs it passes more squat outlier pines, hundreds of yards apart, each with its own stance and shape so that you come to recognise them as aloof individuals, each one with a different mountain backdrop, each one with a view out across their taller and more elegant brother pines clustered in the main wood. The curve and swoop of the wood and the wide, curving crowns of the trees set up the handsome profile of Ben Lui, and if you could train your gaze on that blithe partnership to the exclusion of all else in the landscape you would call the place beautiful. But on every other hillside and in every direction there are com-mercial plantations, hard-edged and unrelieved by either a diversity of tree species or open spaces, except where whole hillsides have been clear-felled. It is old-school forestry, and while in some other places it is being done better now, only the forestry industry ever learned how to make trees ugly. There is nothing wrong with the trees they plant, but there is a great deal wrong with how they are planted, with the brutal nature of how they are harvested, and with the miserable proportion of native trees the plantations accommodate.

I had sought out this high view of the pinewood, for the view over the top of a good wood is often as rewarding for the watcher as it is rare. A couple of hillsides away to the east I know a place to perch in the lee of a broken-down drystane dyke that blunts the west wind and offers a view over the top of a wood where ospreys nested. Aesthetically it is not as good a wood as the Strath Fillan pinewood, for it is a planted Victorian thing of huge exotic firs and rhododendrons, although some of the native trees that the designer of the wood found when he went to work were allowed to keep their place. Birches, willows and alder have nibbled away at its edges, but so have self-sown spruces from a nearby plantation. No matter, the resultant mix is not unpleasant, and because it was laid out with the wellbeing of rich pedestrians in mind, it has its share of good open spaces that must once have been manicured, but now, in its quiet dereliction, nature has moved in with a will. It stands at the mouth of a glen and just below a major watershed with old oakwood remnants to the south, scattered pine and birch to the north, and a lot of commercial forestry. All of these would be ingredients if we ever persuade ourselves to make a new Great Wood. All we would have to do in such a circumstance would be to strengthen the presence of the oaks and the pines, diminish the impact of the commercial forest, and let it swirl all around the Victorian wood, planting young pines and birches in the glen, alders and aspens by the burn, rowans on a few rocky knolls.

The ospreys were encouraged to nest in the Victorian wood after their first abortive attempt in the top of a mobile phone mast. There was a local consensus that an eyrie with a steel ladder up to it was not a particularly sound career move for the birds. So a treetop platform was built and furnished with enough branches to get them interested. They took to it at once, and prospered there a few years until the platforms slipped one spring, spilling the contents of the nest. They hung around for a few desolate weeks, then vanished.

I grew accustomed to walking up to the dyke and sitting there and watching from a discreet distance the male osprey flying in across the treetops carrying a fish for its mate. There was a time when that must have been one of the most common rituals of nature in the wooded Highland landscape. But today, that is the view of ospreys that most people don't see, the long haul that follows the familiar much-televised spectacle of the catch. If you could retrace the whole of that flight, you might see this:

The bird has just heaved itself upward and forward from the water, a ponderous beginning for such a graceful flier. The first few wingbeats are about extracting the bird from water, where it is awkward and ill-at-ease, and into the habitat for which it is supremely designed. As soon as it clears the water and gains the first few feet of height it performs an undignified aerial shudder that frees its plumage of a surprising amount of water. Then the fish has to be secured, preferably in both feet, arranged one behind the other, preferably aligned with the fish head facing forward. The fish is not dead at this point, and is often still thrashing its tail. So head-first is aerodynamically more efficient and avoids a slap in the face by a convulsive fish tail. Some birds are fastidious about this arrangement, others less so, and you do see the odd one holding the fish by the head alone so that it dangles like a limp sock.

The loch has trees and hillsides all around, so the bird must gain height, which is easier said than done, especially if it has bitten off a bit more than it can chew in the matter of the fish. I have heard of – but never witnessed – accounts of ospreys catching a fish that is too heavy to lift from the water, and 'swimming' it ashore using its forewings like paddles in the manner more commonly observed in American bald eagles. The fish is ripped apart on the shore, a chunk of it is eaten, then the osprey heads for the nest with something more manageable in its feet.

The rising bird begins a series of widening circles, gaining height with each lap. One eyrie I know is two miles away from the main source of fish and requires the fish-carrying bird to climb about 1,000 feet before it can cross the intervening range of hills at the lowest point. That's a lot of climbing circles before the bee-line for the distant eyrie tree begins.

If you're lucky, you catch a gleam of sunlight in the white underwings as the bird banks round a hill shoulder half a mile away. Then you lock the glasses on to its flight and you watch it skim the treetops. Sometimes it flies between two of the tallest trees – Douglas firs – and I wonder if that is an example of the bird using trees just as people have done for millennia, as landmarks.

If it has been a long hunting expedition the bird on the nest will start calling while its mate is still a few hundred yards away. A neighbouring mistle thrush occasionally causes consternation because it has learned to mimic that call perfectly, fooling both the sitting bird and the watcher with the binoculars. The home stretch for the osprey is surprisingly indirect. The line of approach often wavers, sometimes stops short where the fish-carrier perches and eats before flying onto the eyrie tree and handing over the rest; even then, there are several passes and circles around and above the eyrie before it lands. The sitting bird's calls acquire a strident edge . . .

If there are still unhatched eggs in the eyrie, the sitting bird takes the fish and heads due north (I have never seen it go any other direction) out across the top of the wood and in a long shallow climb until it reaches a row of disused fenceposts high on the open hillside about half a mile away. There it perches and eats, while its mate settles on the eyrie. Wherever you have a view of a good wood from above, watching such things are possible again and again. You see buzzards, sparrowhawks, woodpeckers, red squirrels, and – if it's the right kind of wood – capercaillie, black grouse, goshawk, even golden eagle. It works best with ospreys because they are such confirmed treetop nesters.

Old habits die hard; I still climb up to the old dyke and look at the osprey nest, scanning the top of the wood in case they have come back and nested elsewhere, but none has shown itself yet. On the other hand, the red kites have moved in, and a wandering young sea eagle turns up from time to time. It will always be a good idea to watch the top of a wood.

I had followed the hill track above Strath Fillan until I got a good view above the pinewood. Here I spent a quiet hour incubating my thoughts with a flask of tea, a pair of binoculars, a camera and a notebook. Then I wanted the embrace of the wood itself. I headed down for the sanctuary of the pines. Crossing the bridge (the lower of the two, the less tricky to negotiate, and the one blessed by a grove of aspens at its east end) is something akin to Alisdair Maclean's Corran Ferry moment, for beyond the far end is a place whose embrace has always felt hospitable, so that I can never quite escape the sense of arriving in the shoes of the prodigal son.

In this of all woodlands that I know, I feel somehow recognised, acknowledged by the very wood itself. I have known it for thirty years, but for five or six of those I lived just a few miles away. It was then that I began to connect with the place in a quite different way; a connection born of intimacy, the sense of a place with an identity as real as my own. This is difficult terrain for a nature writer for it lays him open to charges that vary from 'elitist' to 'precious' to 'holier-than-thou', and I have been called all three. So let me be as honest as I know how. This wood, and a couple of other places within what I think of as my working territory, impressed me in particular because the more I became familiar with them the more I sensed their store of mystery and a kind of natural wisdom that underpinned every aspect of life there. There is an order, a discipline, a rhythm, and it is nature's doing, existing quite outwith the world of people.

None of this dawned on me quickly; there was no bolt of lightning, but rather it grew on me in layers, like moss, over years of going back again and again, working and re-working what I had assumed were the same set of circum-stances, only to find that the details within changed all the time. I reached a point where I became frustrated that so much was out of my reach because I thought I was not good enough at tuning in, not accomplished enough at taking notes and turning them into biological fact, not *nature* enough for nature. But then I grew out of that feeling and grew beyond it to a place where I value the mysteries of nature beyond all else. Nothing in all nature knows all nature, understands all its mysteries. Nothing is immune to the changes nature constantly imposes, but ours is the only species that fights the changes, that upsets nature's order, and the effect of that over time has been to put distance between our species and nature. That distance is what I try to shrink, and the best way I know to try is to work within my territory, to become intimate with a few places, to understand as much as nature is prepared to offer, and to marvel at and treasure the mysteries.

If you were to ask me about the places where I go to strive for that intimacy I would tell you about the pinewood, about a mountain, about a high, alpine-like glen where golden eagles nest, about a reed bed at the head of a loch where swans nest, and about a stretch of slow, alder-lined river and its floodplain. And you can draw a single line that connects all of these places and it will not be longer than 20 miles.

There is also this: the things you learn from such places, the sense of nature's order, the patterns that repeat and slowly evolve, the aloof mysteries . . . all that knowledge is transferable; and if it has been honestly won and if you carry it with you wherever in the wild world you travel, then the same principles apply. And you may be surprised to find that you sit by a strange river in a strange land watching a beaver at work on a cottonwood grove under a strange mountain and suddenly you become aware of that same sense of nature's acknowledgment that you first felt in a pinewood in Strath Fillan 5,000 miles away.

So I greeted the aspens, crossed the bridge, and stepped into the pines, and if you adhere to the philosophy of American architect Frank Lloyd Wright – 'I believe in God but I spell it Nature' – then you might think of such a place as sacred, and you walk more softly than before (although you stop short of taking your boots and socks off, what with the fallen pine cones, the harsh heather and bog myrtle stems, the holly, the sudden bogs, the hidden rocks, the ants . . . especially the ants). I try and empty my mind of everything other than an openness to where I am and what passes here. I climb first to a high narrow ridge that rises above the middle of the wood just to watch the rise and fall of the trees from within. It is also where natural regeneration is most prolific; pines like the dry, well-drained knolls, and the lower, wetter ground is the province of bog, moss, bog myrtle, alder, willow, birch, aspen (that mostly cling to the burnsides). The higher up the corrie, the better the regeneration, a self-evident truth that makes the uphill commercial plantation all the more misconceived. The pines are held in check several hundred feet below what would be their natural treeline.

I like, too, to get down into the hollows so that the trees crowd round and above, and to see the further reaches of the pinewood as skyline tiers. The sense of a truly big pinewood is almost tangible. Oh, but how I long to see pinewoods that disappear round the next hill shoulder and the next, hills robed to the waist in big, domed pines like the ones in this wood. But in this country the sight does not exist; every pinewood is isolated from the next, and nature must make what it can of the fragments. I have seen what a landscape of pine-covered hills looks like in Norway, rolling away from a high mountain summit in green waves, and it took my breath away, even though the pines were shaped more like spruces, not like our broad-crowned trees at all. But there are wolves within those forests, and brown bear, beaver, lynx, moose . . . for it truly is a Great Wood, and any Great Wood in the north of the world needs all of these to be Great.

There is a long, wide heathery slope that climbs at a gentle angle from the heart of the Strath Fillan pinewood up to its western flank. The trees line both its edges but none grows out on the slope itself so that it looks as if nature has created a parade ground for the staging of great events. A small army of people could march up there 20 or 30 abreast. I found the slope one sunny evening with the heather at its purplest and the sun streaming down it from the west, the whole wood basked in a benevolent glow. I leaned against a birch near the foot of the slope to be part of its suddenly rarefied atmosphere. I had been still there for perhaps half an hour when a red deer stag stepped from the trees, walked out into the middle of the slope and then turned at right angles and started walking uphill towards the skyline. It was August and his condition was prime; his coat was glossy, his head was high, his antlers ten-pointed and wide, and his feet kicked up tiny clouds of purple dust. Then a second stag appeared, following, made the same turn at the same place and there were two stags heading for the skyline. Then a third, then one at a time and several seconds apart, there came 15 more. They walked in single file at a quiet, measured pace until the first stag was almost at the skyline. Then he stopped.

The others walked up to him and stopped too, not in single file now but tightly bunched and suddenly nervous. There was a lot of movement in the group, but it remained tightly gathered. I could hear the muffled thud of restless feet on the hard, dry ground, the occasional rasp of antlers touching. Then in the gap in the trees and directly over the stags' heads the silhouette of a golden eagle appeared. It circled once beyond the stags then came in low. The herd scattered and ran. The eagle charged down the slope not six feet off the ground then suddenly appeared to stand on its tail and soared almost vertically for two or three hundred feet, banked high above the trees and headed west whence it had come. Not one stag showed on the open ground.

'And what was that all about?' I muttered to the tree where I leaned.

I have heard of eagles charging deer to try and drive them over a cliff, and seen one memorable film clip of such an incident in the Cairngorms, but there was no cliff here.

These were all mature stags and far beyond the scope of an eagle as living prey; it would have taken a wolf to bring one down. But the eagle knew that. It also knew that safety for the deer was only 20 yards away to left and right over easy ground. So was it making mischief? Was it making a point: I am a threat and in the right circumstances I can drive you off a cliff and kill you? Was it playing?

Every good wood is the stage for moments of drama like these. Mostly, only the trees bear witness.

I wandered away, deeper into the wood where late sunlight lit one side of a particularly slim and tall pine from top to bottom – only that tree in all of the wood that I could see from where I stood. But the sun also caught little highlights of grass and mounds and openings in a way that revealed something I had never noticed there before – a faint, long-overgrown path through the wood. It ran for perhaps 200 yards then disappeared among trees before a cloud crossed the sun and the sense of that path just vanished. It was as if a door had opened, as if the past lay behind it, as if the door closed again. I have since stood in the same place several times and tried to recreate the trick and failed. But it was there, and I photographed it and the ghost of an old, old way through the trees was fleetingly tangible. Every good wood is the stage for moments like these too, and mostly, only the trees bear witness.

The really surprising thing about the Coille Coire Chuilc is that although even a thousand years ago it must have been part of a forest that stretched for many miles, what has survived feels so intact. It has the feel of a big wood, with clearings, with dense regeneration, with individual trees of great stature, with standing dead trees (one of which unleashed two great-spotted woodpeckers at me in a low and direct flight that almost parted my hair), and a vigorous understorey. One of the most striking of all its trees is also probably the tallest, and this despite the fact that at ten or twelve feet up, the single trunk suddenly bends at an angle of about thirty degrees from the vertical then climbs dead straight to the crown at that angle, so that the tree looks as if it is perpetually in the act of falling over. Another has the profile and beautiful symmetry of a great oak.

Water on the move is a constant background sound. The burn that flows down the eastern flank is joined by a second that bisects the top half of the wood. These flow boisterously past the aspens and under the bridge to feed a fast and fluent river that effectively marks the wood's northern boundary. Its banks are mostly steep and rocky, the rocks cut into horizontal ledges where both aspens and pines have rooted (and some have jumped the river to root in the rocks of the far bank). Here in fissures filled with moss a pine seedling has found enough sustenance to produce three six-inch-long stems, and this too, if it is spared the attentions of deer and spates and the worst excesses of the ill-luck that decided its seed's final destination will grow into one of the wood's great individualists. It won't grow tall, but it will become a landmark, for it has a brother directly opposite on the far bank already about two feet tall. I can see them in 50 years standing 20 feet high and slim like gateposts, and other pinewood wanderers will mark them in passing with an affectionate nod of familiarity, as I do myself even now.

These are better times for this wood than they have known for a long time. In my own lifetime I have seen the grazing pressures eased, for you used to find sheep wreaking their havoc all over the wood, and the benefits are shown in regeneration, and in the depths of the understorey. There are new trees marching down towards the river from the north, all the way from the main road, the fruits of a community planting project which, remarkably, has taken its cue from the pinewood rather than the commercial plantations that surround the village of Tyndrum at almost every turn. Thousands of native trees now swarm around stands of old birch and a few oaks that used to look only forlorn. Now they look like pied pipers gathering the young of their own kind in ever greater numbers. The result is a matrix designed by optimists who give the impression at least, that the Great Wood is a concept not just with a mysterious past but also a realisable future. And this is what it will look like, what it will have to look like – a tapestry of distinct woodland groups of different ages and species which will evolve over centuries into a blend that nature can work with and people can live with much more easily and willingly than we do with the grubby thumbprints of the forest industry for which neither we nor nature much care.

CHAPTER SEVEN

Glen Orchy and Rannoch

G len Orchy is the centre. There is the glen itself which ferries the handsome but short-lived River Orchy from Loch Tulla to Loch Awe. There is Glen Orchy and Inishail, the historic name for a loosely defined sometime heartland of MacGregor country that extended from Loch Awe in the south (where the island of Inishail is a burial ground of rare distinction whose inmates include several dukes of Argyll) to Loch Etive in the west, and to the Blackmount in the north. Its influence, according to some, was felt all across Breadalbane as far east as Loch Tay and the Black Wood of Rannoch. According to some. You can never be sure when it comes to the historical record of the MacGregors, for history was rarely kind to them and historians have tended to vent their spleen on them or to sentimentalise them beyond all recognition. But their clan crest is a pine tree – a fallen pine tree with a crown lodged in the branches – and this particular reading of that particular tract of country begins with that persuasive symbol. It turns out that the pines of Glen Orchy have been falling for a long, long time.

Whenever and wherever historians and other scribes have turned their attention to the Great Wood of Caledon, and with, it should be said, greater or lesser degrees of credibility, much as others of their kind dealt with the MacGregors, Glen Orchy rears its head. Its name crops up so often that I think that here, surely, was a forest to be reckoned with, one famed in particular for the size and quality of its Scots pine and oak. And I have wondered off and on over the years . . . was *this* the Great Wood, not a cloak of all Highland Scotland but a slung barrier that defended the Highlands from west to east? The Romans, if indeed it was the Romans who gave voice to the notion of the Great Wood, were accustomed to consolidating their gains and defending them by slinging barriers of their own making from one side of the country to the other. Could it be that this wall of trees was what finally turned them back, the westernmost extremity of their empire, a barrier they judged to be beyond them?

I have been accustomed to trekking the hills around Glen Orchy and camping by its river for many years, and long before the notion of the Great Wood began to get under my skin. But in the 50 or so years since my father drove us – his wife and two sons – through the glen during a family holiday based at Benderloch on the Argyll coast and I fell for the place in a quite unreasoning and uncritical way, it has metamorphosed from a butterfly into a slug. It has the forestry industry to thank for this unkindness. Given the historical standing of its forest, few places the length and breadth of the Highlands are less deserving of the kind of blunt-instrument forestry that characterises today's glen.

Only the south-west end of the glen has been spared; some huge oaks grow by the road, and some of the south-facing slopes are lightly wooded with oak in particular, occasionally to a surprising height. From beginning to end the river flows between banks lined with alder. Beyond that native fringe almost nothing relieves the overwhelming suffocation of spruce. Yet still, wherever they can grab even the most flimsy toehold, usually on narrow patches of land between the road and the river, the old order still shows its face – pockets of aspen, handfuls of Scots pine, a few willows, birches absolutely everywhere they think they can get away with it, rowans, a stand of big ash, a few hazels. Alders sneak away from the river up the steep, dark banks of almost every burn.

In Glen Strae west of and parallel to Glen Orchy, there is a burn called Allt nan Giubhas, and an Inbhir nan-Giubhas marks the place where it joined the River Strae. Allt nan Giubhas is the Burn of the Scots Pinewood and Inbhir nan-Giubhas is the Mouth of the Burn of the Scots Pinewood. River names are the oldest on the map, and almost invariably mean what they say. The landmark survey by Steven and Carlisle, *The Native Pinewoods of Scotland*, found puny remnants of living pines in Glen Orchy and Glen Strae but much evidence of substantial old lost woodlands. Steven and Carlisle make the point that because native pinewoods reach back to the ice age in 30 generations, they are 'not the least of our historical monuments'. The pinewood at Strath Fillan five miles back down the road (and surely once a component part of the greater Glen Orchy Forest) confirms that judgment.

Yet comfortably within living memory – or as it has turned out, uncomfortably within living memory – the beautiful Highland glen I was introduced to at the age of 11 or 12 has been sacrificed to the timber industry without a moment's thought for aesthetic sensitivity or considerations of either natural or human history. In 1959, which is roughly when I first saw the place, it must have looked very much like Glen Finglas did in 1996 when the Woodland Trust bought it. But instead of an enlightened mission of rescue and recovery, Glen Orchy has suffered an entirely opposite fate, and it is doomed probably for decades to a relentless cycle of monoculture crops and clear-felling. I can see nothing to stop it unless the Scottish government dares to reinvent the Forestry Commission and reorder its priorities.

Aesthetics matter. Planting new woods or restoring old ones changes the character of the landscape more comprehensively than any other influence we can bring to bear on it. Well designed planting that takes its cue from nature with a wide range of species creates all manner of new opportunities, and nature can be relied upon to take full advantage of them. It creates new opportunities for people too because it is labour-intensive; the work is reliable and produces a skilled workforce with an attachment to the land. What has been inflicted on so much of Glen Orchy by current forestry practice denies opportunities for nature by smothering the land and rejecting diversity. It denies opportunities for people because the work is done by imported squads that mass produce forests, and creates a relatively unskilled and disinterested workforce.

But this is not a new phenomenon. A local minister, writing in 1843, lamented the demise of Glen Orchy's woods: 'Not so long ago, the greater part of our moors and valleys and the sides of our mountains, midway to their summits, were clothed with trees of various kinds. The braes were clothed with a dense and magnificent forest, partly of oak, birch, ash, and alder, but chiefly of pine . . . The hills are still partially clothed with oak coppice, birch, aspen, elm and holly, but these generally speaking are rapidly disappearing and our mountains and valleys and straths have become comparatively naked and bare.'

By that time the Highland Clearances were in full swing, the Highland people were being replaced by lowland sheep, and as far as the minister was concerned the rot was well set in for the local woodlands, and it has not stopped. But the minister's assessment was delivered in the midst of a period of reckless deforestation that had been accelerating for a couple of hundred years, and which achieved a spectacular nadir in the 1720s and '30s. The earl of Breadalbane had negotiated a deal with an Irish consortium in 1723, the kind of relationship that had become fashionable by that time, especially in the West Highlands. Ireland was critically bereft of timber and tanbark so squads of men formed a loose partnership and crossed the sea to buy standing timber from landowners, fell it, and transport it back home.

This particular group of chancers quickly fell foul of the local people who considered their long-standing practice of making use of local timber to be an inalienable right. The deal between the hard-up Breadalbane who desperately needed the cash, and the Irishmen who desperately needed the timber, did not seem to have paused for a moment to consider that right.

Now Breadalbane had a fair old fiefdom to oversee and when he did deign to visit it he preferred his Loch Tayside residence to Glen Orchy, and he preferred his Edinburgh residence to Loch Tayside. Two years into the contract he went to inspect the fruits of the Irishmen's labour and was aghast to discover that 'they have not left one standing oak tree in the country', and were now laying waste to his precious pinewoods. The contract had been poorly nego-tiated, and the Irishmen cut swathes through it, and continued doing so for about 15 years. 'Not the least of our historical monuments' were sailing across the Irish Sea in their countless thousands.

For many years, my favourite destination in Glen Orchy was a plateau-topped mountain called Beinn Udlaidh. It means the Dark Hill, presumably because it was named from a community in the glen that was accustomed to seeing it against the light of a southern sky, for in every other aspect my earliest memories of it are as a place of unfettered light. The long quartz rib that defines the upper reaches of its broad north-thrusting ridge is visible for miles, glittering in sunlight, and accommodates a high, secret pool of the clearest, coldest, sweetest water I have ever seen or felt or tasted. The darkening forces came later, the first tree hordes that swarmed over its eastern flank and round the northern prow, then a new hill road was carved up to the very rim of its handsome northern corrie and the invasion of trees unfurled a second wave.

I don't like writing about trees in terms of hordes and invasion, for trees are the most benevolent of all nature's flora, the most generous life-givers, the most hospitable shelters, but trees like these, planted like that, achieve the precise opposite of benevolence, generosity and hospitality. To cover a hillside in trees that repel, that shun wildlife, that create a non-habitat, and all this in a glen that was the centrepiece of an ancient forest, is a grim and unworthy response to a beautiful landscape and to a responsibility to honour the history of that landscape. These first new plantations were the advance guard of an occupation of all but the south-west extremity of Glen Orchy, and really nothing excuses such brutality. The place is wrecked.

Again, when I first began to climb in the glen, I used to daydream about living in a small white cottage that sits quite on its own beneath the mountain. Its name is Invergaunan, which is a crudely Anglicised rendering of Inbhir Ghamhnain, the mouth of the nearby burn, Allt Ghamhnain. The name also adheres to the corrie where the burn rises, and the short glen it flows through on its way to the River Orchy – Coire Ghamhnain. My less than conversational Gaelic guesses the name is something to do with young cattle, a notion which is at least sustainable given the presence of shielings in the glen. At any rate, this glen too was wooded, for over the years I have unearthed a few bleached pine remnants in the peatbogs, headstones to long-dead trees, the oldest inhabitants of the place.

A hundred yards away from the front door of Invergaunan, a ruinous drystane dyke is attended by a stand of half a dozen Scots pines. There was a time in my much younger days when I ached for the bare hillsides of this place, and I romanticised the heroic stance of the pines amid the stones, greeting them as the last survivors of a lost warrior race. Now as that tsunami of commercial forestry devastates one hillside after another, obliterating everything in its path, leaving almost nothing alive but a bleak press of trees that were never intended by nature to grow that way . . . now I bear witness in a quite different spirit.

Song for the Pines

Stand while you may,
while I pray to God knows
what manner of deity, seeking
the reconsideration of this bleak design
that weakens Nature to its knees,
and wondering aloud
that if there really are immortals among trees
then may that stand of old Scots pines be these?

I think it might take the immortality of Glen Orchy's handful of Scots pines and armful of oaks to turn aside the forces of darkness ranged against the natural woodland order. I look at that one-verse poem and acknowledge at once the desperate nature of its straw-clutch. Grant these trees immortality and then wait for the madness of the worst excesses of commercial forestry to abate into a kindlier era when they might be thinned and opened out to accommodate planted descendants of the immortal ones. It is, as any professional forester will doubtless tell you, a less than plausible strategy to submit to the inner sanctum of the Forestry Commission, a nature poet's response to the real world rather than a scientist's, or for that matter, a forester's. I would argue that the real world is the one that nature laid down, and his is its unreal usurper. But he has a point in that, as far as I know, no tree has the capacity to tune in to a nature writer's proffered reassurances, no matter how sincerely voiced. There again, that is only as far as I know, and there is a great deal about the art of communication between mankind and nature that I don't know or understand, but none of that stops me from trying.

So I have been known to sit in the midst of that little gaggle of Glen Orchy pines I have known for more than half my life, the centre of the centre as I see it, and to urge the patience of immortals on them – 30 generations back to the ice age – because I still believe (despite the evidence of my eyes) that Glen Orchy – the greater historical Glen Orchy and Inishail from Loch Awe to Rannoch – can recover, can be healed; still believe that there are better ways of both growing timber and honouring landscape and the spirit of trees; still believe that I am far from a lone voice, that there are others out there who preach the same gospel to a small congregation of Brother Pines. Once, I met such a kindred spirit in a small pinewood not far from here.

*

It was a soft day when the light was flat, the air moist, the sun absent, the pines emitting that melodramatic blue-green luminosity that strong sunlight would obliterate. The same flat light strikes unsuspected notes in the pine bark. The range of grey-blues, browns and reds compounded by lichens and jigsaw-piece shapes and shadows that characterise the bark can startle a seeing eye. The older the tree the more elaborate the woody mosaic. And when I first saw her she was being a piece of a tree.

It was almost certainly the oldest tree in the wood, and I had been watching it from a distance before I realised that there was more to its great trunk than just trunk. Between five and six feet up from the ground there was a dark stain, but I could make nothing of it and it began to puzzle me. I put the binoculars on it – normally you don't need binoculars to look at a tree – and realised at once that it was the long dark hair of the back of a human head. The rest of the human in question, I now realised, was dressed in clothes that so resembled the patterns and shades of the tree trunk that it was all but invisible. Now that its cover was blown I could see that the face was turned inwards to face the trunk and the arms were spread wide in embrace. Then the head turned sideways to put a cheek to the tree and in the same movement put me – with raised binoculars – in its field of vision.

She let go of the tree and turned to face me. I lowered the glasses. It was the least I could do.

'Hello,' I said.

'Hello.'

The silence that followed was not so much awkward as

110

wondering. We were both under the spell of trees but neither of us had any idea what the other was up to. I indicated my small backpack:

'I've got tea.'

'Oh. Ah . . . thanks.'

She came over, moving like a fox, I thought, or maybe a wolf, but she was dressed like a tree. It was one of those flasks with two tops that make two cups. She nodded her thanks. I said:

'Funny how flask tea always tastes more like flasks than tea.'

'I didn't see you,' she said.

'I didn't see you either – your clothes.'

I saw now that she wore a dull green jacket not unlike my own except that hers was smothered in hand-painted pine-bark patterns. Her trousers and her shirt matched.

'Tree bark,' she said, as if that explained everything. Then:

'You're a stranger in my wood, watching me through binoculars. Why would I explain anything to you?'

'No. You're a stranger in my wood.'

'You mean you own this wood?'

The thrust chin and the dark challenging eyes and the tone of voice matched, the way the shirt matched the jacket that matched the trousers that matched the trees.

'No. And you obviously don't either.'

'Why do you come here?'

'I'm a nature writer. I look at trees and whatever else keeps them company. What's your excuse?'

Her first smile. Good smile.

'Two excuses. One is to save young trees.'

Silence then, awkward this time, and reluctant.

'The other?'

'To be a piece of that tree.'

'A *piece* of it?'

She shrugged, clammed up. How could I understand? Then she relented. She was drawn to trees. Some people saved stray cats, or whales, or white rhinos, or giant pandas. She saved trees. Scots pine trees. She had a small piece of land, a small sphere of influence. Not much, about this size, she said with a gesture of outstretched arms which may have meant the clearing or the whole wood. She came to woods like this one where the young trees had no chance . . . the deer, the sheep, the indifference of the estate . . . then she stopped, frowned, and started again.

Did I know the Botanic Gardens in Edinburgh? I did. There was a big Scots pine, a giant, over where the Henry Moores used to be. I remembered. Reclining bronze hermaphrodites, perplexing holed curves, solid holes, hard suggestive softness – *how did he do that?* It blew down in a storm. She saw it a few days later lying prone. *Reclining Pine Tree – Moore should have made that.* She walked up to it and put her arms round it. That was when she realised it was not yet dead. Her eyes filled, dark pools, and she said:

'She filled me with her sadness.'

'She?' She nodded.

'The most colossal sadness.'

'What did you do then?'

'This.' She opened a small shoulder bag. It was full of pine seedlings. Surely she wasn't planting young pines here where they would be bitten to death, as sure as there will be midges in August?'

'Not planting them. Stealing them. I plant them where I know they will grow. They would die here. I can't help this wood, but it will help pine trees. It will help assuage the great sadness. I feel called to it.'

'Called?'

'Yes. An example to other people.'

'Like who?'

She looked at me directly then, the first time I felt she was addressing me directly rather than the surrounding trees.

'Like you.'

'But I'm converted. I already come here for the trees.'

'No, you come here for what's left. I come for what will be. Perhaps you will write something. Perhaps someone will read it and do this somewhere else. Move trees to where they will be safe. I can do nothing for this wood. I have no sphere of influence.'

'But you care for the old trees too. You were hugging the old one when I saw you. It's not dying. I presume it's not sad.'

'She mourns. The lost generations. She shows me her grief. I was not hugging her. I was becoming a part of her. I can do that. I can reach her. She remembers wolves.'

I looked at the tree that remembers wolves, then back at her.

'*She* remembers. Always *she*.'

The woman nodded.

'She is my sister.'

*

North of the glen an unpromising little road turns its back on Bridge of Orchy's two landmark mountains – Beinn Dorain and Beinn an Dothaidh – and burrows west, its destination the lonely inn at Inveroran and the savage beauties of Loch Tulla. It is a wild place of big winds, a good deal of rain, heavy winter snow and numbing cold, and the tumultuous mountain panorama of the Black Mount. This was the birthplace in 1724 of the greatest of all Gaelic poets, Donnchadh Ban Mac-an-t'Saoir – Duncan Ban MacIntyre – fondly remembered to this day as Fair Duncan of the Songs. His greatest achievement, 'In Praise of Beinn Dorain', was a 550-line masterpiece meticulously constructed around the rigid formalities of a pibroch, a grand and uncompromising gesture in a landscape of grand and uncompromising gestures, and it is one more set piece in the elusive jigsaw of the Great Wood. Standing by the ruinous cottage that was his birthplace I was thinking of Alasdair MacMhaigstir Alasdair and his 'Song of Summer' that summoned the *feadan* to his cause, and wondered again about the relationship between pipe music and the Great Wood, and of the pipemaker who loved yew but didn't like to inhale its dust, and then I thought how in North America the Sitka spruce is the choice wood for many of the best guitar makers and here it is pulped to make hardboard.

Late in the afternoon I was wandering through the pinewood on the south shore of Loch Tulla, eyeing the comparatively extensive wood on the far shore, when my binoculars drifted past two birds that gleamed vivid white in the hard and sunless afternoon light of late March. It was cold, the loch and mountains were a forlorn grey, there was fresh snow on the summits of the higher hills, and nothing about the scene said spring at all. But I steadied the binoculars on a low limb of the nearest pine and watched the two white bird shapes until I could make sense of them, which is another way of saying until I could confirm my suspicions. They were far across the loch and only that gleam of white gave them away where they drifted close together on the water. Then I realised they were dozing; they had their heads turned back to face their tails, and that delayed the confirmation process. Then a scattered flotilla of goldeneye purred across the loch in flight, and their passing disturbed the drifting pair so they raised their heads and fully righted their bodies and became what I thought – hoped – they might be, which was a pair of black-throated divers.

Then, having stirred themselves, they gave voice. If you have not heard the duet of a pair of divers in a setting like this you have not heard the national anthem of our wilderness. If nature had invented the pibroch it would have sounded like this, and perhaps nature did and perhaps a man with music in his head a few centuries before Alasdair and Donnchadh listened to two divers newly arrived from their winter waters in the east coast firths, and made the first attempt with the *feadan* in his hands to make music of nature.

A small moral of the story is that open water is – was – part of the Great Wood too, and the hordes of forest grazers came down to the shores of lochs like Tulla to drink in the evenings and to cool off in the brief heatwaves of such a land. Watersheets affect the life in the forest too, and reflect

115

it on the still days so that your eye never falls on a familiar landscape and sees it twice in exactly the same way. And the big lochs like this one, or Loch Tay or Loch Ness, were accustomed to freezing over from end to end in colder eras than ours, allowing the wolves to use them as highways. When this one froze it was a land bridge between Glen Orchy and Rannoch.

Even today, what's left of the pinewoods helps you to navigate back through a few millennia, so I sat on the shore among lightly scattered trees with a thick and climbing pinewood behind me and the long sprawl of the pinewoods of Victoria Bridge on the far shore. The divers called with voices that cut the cold air like sabres with a haunting downturn at the end, and I made what I could of the old forest of Glen Orchy and Inishail. The pinewood at Strath Fillan lies at the foot of the north-facing slopes of the first great mass of Highland mountains if you come up through the middle of the country from the south. There is another, smaller wood at Glen Falloch at the north end of Loch Lomond. And these belong to each other. Going east, there is Glen Dochart of the oaks then Loch Tay, whose southern shore in particular continues the theme.

Glen Orchy is the centre, a power base of nature and people, a Great Wood of which Glen Falloch, Strath Fillan, Glen Dochart and Loch Tay were a southern rampart. Most of the glens as far west as Loch Etive and the top of Loch Awe were wooded, and bear old names that prove it. Loch Tulla was surely surrounded by pinewoods, and from there and all the way to the Black Wood of Rannoch – another miraculous survival of the old order – there was a single forest, with, at its heart, the high and primitive wilderness of Rannoch Moor. It is tempting to look at Rannoch Moor in its beleaguered 21st-century condition and scoff at the very idea of a forest. But trees are everywhere on the moor, or at least the bones of them are. Carbon dating puts many of them in the second millennium BC, yet Fraser Darling speculated that it was still wooded when the Romans came 2,000 years later. The authors of *A History of the Native Woodlands of Scotland* point out that something catastrophic wiped out almost all of our Scots pine around 4,000 years ago, and I wonder if Rannoch Moor was a victim of that climatic onslaught and simply never recovered, perhaps because of the extreme nature of its climate.

The moor was pivotal in the shaping of the south Central Highlands, a towering ice cap that spawned the glaciers that carved the glens and defined the mountains of all that land, the last ice to melt and so the coldest, wettest land least equipped to sustain good trees.

I suspect that all this was the greatest tract of the Great Wood, that it was wilder and much more pine-dominated than the oak-and-birch-laden woods of Menteith, which we are now inclined to call the Trossachs. In *The Last Wolf* I voiced the argument for the reintroduction of wolves into Highland Scotland. I chose Rannoch as the first release site because the largely undisturbed land of the Black Wood, the Moor, and Loch Tulla is of an uncompromising wilderness character that matches the wolf's, and because being a part of that greater forest of Glen Orchy and Inishail and Breadalbane, it lies at the centre of the Great Wood itself.

That in turn offers an emergent population of wolves the opportunity to expand both north and south if it chooses. An ancient darkness would be banished, for the wolf is a catalyst, an enabler, a creature that creates endless opportunities for nature in all its guises, and like aconites appearing through frozen ground they scatter the land with new points of light.

Rothiemurchus

O
ld books say the name means The Broad Plain of the Firs. Modern revisionists say that the phonetics invalidate that idea and that it derives from Rat Mhurchais, The Fort of Murchas. I side with the old books.

The long-forgotten Fort of Murchas was far too inconsequential a footnote to the history of Strathspey and the northern Cairngorms to have given its name to such a defining tract of ancient woodland and mountain. This is a landscape named for its natural features. The more obviously translatable Gaelic hill names include the Hazel Hill, the Yellow Hill, the Promontory of the Pass, the Black Crag, the Black Peak, the Pale Hill. There is even a Hill of the Clump of Pine Trees. If the ancient namers of the landscape thought it appropriate to single out a group of trees to identify a hill, it is hardly likely they would have ignored the massive impact that the Scots pine forest of

Rothiemurchus makes on its landscape setting when viewed from even the slightest of elevations, nor for that matter the psychological impact it must have made on the earliest travellers. The Broad Plain of the Firs it surely is.

They used the word 'firs' in eighteenth-century books when it was still a generic name for conifer trees. In the case of Rothiemurchus, they meant Scots pines. What Seton Gordon called the Scots fir in the early twentieth century is what we now call the Scots pine, and in Rothiemurchus the Scots pine is an omnipresence. If you walk the Gleann Einich track from Coylumbridge you are immersed almost at once in a *depth* of trees such as you will not encounter anywhere else in Scotland – trees to darken a sunny day.

I cannot walk here without thinking of bagpipe music, and with pibroch in particular. I made the association in my head long before I knew what underpinned Fair Duncan's 'Beinn Dorain', and much longer before I had even heard of Alasdair MacMhaigstir Alasdair. The pines hold the ground – the *urlar* – and the variations, the birches are the grace notes, while the bass drone is the work of junipers that stand around under the pines like bloated, foggy-fleeced, grey-green sheep.

All these crowd round. An atmosphere of trees bears down. You look left and right and at first all that happens is that the forest moves past you, tree by tree by tree by tree. You hear your own feet, your own breathing, and these move to the rhythm of the pibroch in your head.

A foot stamps.

You startle, whirl towards the sound, freeze.

What the . . .?

At the end of your gaze a roe doe is watching you from between two junipers 20 feet away. You remember to breathe out. She is unafraid. You relax and then you enjoy her. You take two photographs, five minutes apart. Later study of the images will teach you that in those five minutes the only part of her that moved was one of her ears. She was trying to tell you something about the worth of stillness in the company of nature, in the company of trees.

You walk on and the pibroch resumes. You try and walk more slowly. You remember the piper who showed you how to do this, his slow, slow march, his timing, his timelessness. Time was, that roe deer could have been a wolf curiously watching your passage, head on one side. Twenty or thirty years hence, such a time could return. Oh, we should be so lucky!

Sunlight and wind are up in the canopy of the pines. Down here they may briefly dance around you like butter-flies, the pibroch with a jig in its step, then abruptly all is measured and woody again and your own footfall is the loudest sound in the world.

About the piper. His name is Finlay Macrae, he is a Skyeman, a Mod gold medallist (which places him among the hierarchy of pipers), and once I heard him play in a pinewood. It was not here in Rothiemurchus, but in Glen Affric west of Loch Ness, and if there is a single pinewood, a single souvenir of the Great Wood anywhere in the land that might justly claim to rival Rothiemurchus in grandeur, it could be Glen Affric. Finlay Macrae's day job was that of a forester, and sometime in the 1970s he had won an international award for his restoration work among the pines of Glen Affric. I was a newspaper journalist at the time, and just beginning to flex my muscles as a writer about the natural world. Shortly before his award I had written something critical of the Forestry Commission – Finlay Macrae's employer at the time – in the *Glasgow Herald*. I was invited to join a party of journalists to meet Finlay Macrae in Glen Affric.

We duly met on a Friday evening in a hotel at Cannich. There was a ceilidh and whisky and he played the pipes, and although I was no authority I know a good musician when I hear one and a very good one when I'm standing in the same room.

The next day we went out to the pinewood in Glen Affric and he played the pipes again, but this time he played pibroch, and his slow, slow march as he played had something of the strut and the omnipotence of the cock capercaillie about it. The music rose and hung in long sinuous clouds on the still air of the morning and eddied among the pine trees, and its pulse was like golden eagle wingbeats. Suddenly I wanted to be alone with the music and the trees and to try and fathom why it was they belonged so utterly each to the company of the other; that morning I formed the habit of a lifetime which is to put landscape and music together in my head and make of them something indivisible. Such was the mastery of the piper, such was the piper's love of the landscape where he played that it infused and further elevated the music, such was the power of the landscape. It was the first time in my life that music moved me to tears.

So pibroch became the voice of great pinewoods, and

because I was born in the east of the country in Dundee and found my way into the wild places of the land through the native's tradition (a northwards progression by way of the Sidlaw Hills, the Angus Glens to the promised land of the Cairngorms), Rothiemurchus has always reigned supreme in the hierarchy of Highland forests and of pinewoods in particular. So I packed something of Finlay Macrae's music when the weekend was over and it has been my good companion ever since.

Rothiemurchus is unique among the landscapes of Highland Scotland. I have known it for more than 40 years, and I have walked the Gleann Einich track for as long, and always, always, that sudden and so-deep immersion takes me by surprise. It is a bit like high-altitude climbing in that it requires acclimatisation, demands of me that I walk more slowly, have more care about how and where I place my feet, think more deeply about where I am and what I am trying to do here. Perhaps if I lived nearby and walked here two or three times a week instead of once or twice a year I might accustom myself to it, but even then I am not at all sure. Nan Shepherd, poet of these mountains, wrote that 'the thing to be known grows with the knowing' and if there is a single philosophy more appropriate to this of all landscapes I cannot imagine what it might be.

The first time I came here I must have been about 19 or 20, so all of ten years before Glen Affric. I was besotted by nature, by wildness, and energised by my first encounter with *Ring of Bright Water*. Gavin Maxwell's book redefined what my relationship with the natural world would become. I had read it and thought: 'I want to do that.' I didn't want to live with otters in my living room and my bed, but however my relationship with nature panned out, I knew that I wanted to write it down. The height of my ambition became that I should write my own *Ring of Bright Water*. That was the mindset of the young man who took to the Gleann Einich track for the first time, somewhere around 1967.

Almost inevitably, I romanticised the Highlands then. Almost inevitably, I romanticised Rothiemurchus. My first impression was that everything was ancient. This was surely the living, breathing Great Wood of Caledon which I had read a bit about in books and magazine articles by the likes of Seton Gordon, Tom Weir, and Frank Fraser Darling who, you may remember, had written of it thus: 'the imagination of a naturalist can conjure up a picture of what the great forest was like: the present writer is inclined to look upon it as his idea of heaven.' So elevated was Fraser Darling's reputation at the time that he persuaded me among many, many others to be likewise inclined. So if Rothiemurchus was heaven, then these trees were surely ancient and immortal and the fruits of a celestial lineage.

What I know now that I did not know then is that Rothiemurchus is a much ransacked pinewood, that it was described as 'exhausted' in the mid nineteenth century, that its red deer had been hunted to extinction by the beginning of the nineteenth century, that it was turned into a deer forest in 1843 thanks to a scheme to re-stock it with animals from the Mar estate, that its timber was plundered again to feed the rampant appetite of the twentieth century's two world wars, and that when it was declared a national nature reserve in 1954 the designation was as much to safeguard and fund the improvements that future conservation efforts might achieve as it was in recognition of what it was at the time and what it may once have been. Long before any of these eras, of course, it had lost all its big mammals, the wolf being the last to go (probably later than we think, but also probably somewhere around the end of the eighteenth century), and many of its characteristic bird species, notably the capercaillie, the great stuff-strutting grouse of the pinewood (shot to extinction around the same time as the wolf). What the Rothiemurchus of 1967 was trying to tell me, and what the subsequent decades of conservation and restoration confirm, is that given half a chance it is capable of endless renewal and healing. Because if the Great Wood of Caledon is anything at all, it is a continuum.

Some of the characteristic elements of the forest simply go on and on, endlessly renewing an ancient pattern. The first among these is the golden eagle, which survived even the Victorian killing fields when a hooked beak was practically an invitation to slaughter. Naturalist and golden eagle authority Roy Dennis explored an intriguing theory in a contribution to *Rothiemurchus: Nature and People on a Highland Estate, 1500–2000*: 'I actually think that some landowners just liked golden eagles because they were involved in their history, in their relationship with the land, going right back into time. It is a great pleasure that the eagles are still here. Their nesting eyries are historic places, the same places being used for centuries. The nesting sites on Rothiemurchus which I first studied in the 1960s were the very same that Seton Gordon visited in the early years of the century.'

Some of these very nesting sites are in pine trees, for here and there in the eastern Highlands, the golden eagle is a tree nester. Seton Gordon once described a tree where golden eagles had nested to his certain knowledge for over 50 years. The tree contained several eyries and had been completely reshaped by the eagles. In *The Golden Eagle* (1955), the second of his eagle books, he described an eyrie in the crown of a 'Scots fir' at 1,800 feet: 'Although the tree was old, it was of no great size and the eyrie was so large that it covered the whole of the crown of the tree. The nest was nearly eight feet in diameter, and its depth may be judged by the fact that when I climbed the tree and stood (I am six feet one inch in height) on a branch at the base of the nest I was unable to see into the cup . . .'

The solitary observations of the naturalist and nature writer go on, and they too have their place in the pageant, gathering knowledge and – hopefully – wisdom, and furthering the understanding of our own species. So when I walked the Gleann Einich track for the first time my footprints were the first in my personal history among these distant echoes of the Great Wood. Two or three years later, I went to the Sleat peninsula of Skye for the first time, and together these two landscapes – and the literature they had spawned – became the foundation stones on which I would eventually build a writing life. They taught me the worth of reworking the same landscape circumstances again and again, comparing notes, making judgments, talking to older, wiser heads with wisdom to impart, and questioning everything I had not seen and learned for myself. And when I finally became a nature writer for a living in 1988, I had already established a core working territory on my own doorstep in the southern Highlands with Loch Lubnaig and Balquhidder at its centre. There I was able to practise on a daily basis the lessons I had learned on Skye, and especially among the pines and mountains of Rothiemurchus, one more graduate to emerge blinking in the sunlight from the shadowy halls of the Great Wood.

The track burrows deeper, creeps beneath the densest part of the forest then begins to climb. You become aware as you climb with it that you crave air, light, a breeze on your face, sun on your face. Sure enough, the façade of trees begins to crack, then to break apart. There are fissures in the canopy, then spaces, then quite suddenly the track is at the edge of the wood rather than the heart. There are still trees to your left but a void has opened to your right, a thing of space and light and sunlight, a gently climbing slope of heather, grass, rock, bog. There is no denying the relief. But then you notice that there are trees above and beyond the space, then that even this space nurtures trees – a thin and twiggy smattering of saplings from knee-high to head-high, some old and weary and withering, some new and green and eager, some birches in the wet bits that don't appeal to pines, some small clumps of juniper like tumbleweeds that have run out of steam and simply stopped tumbling.

The first lesson: proper forests, proper Highland forests, are not endless and impassable, they are open, they harbour different densities of trees, their understorey varies, they like clearings, their resident mammals demand clearings; at any one time, any time at all in the entire history of the Great Wood, after say the first few post-glacial centuries, there is every age of tree from inches-high shoots to hoary old veterans of 70 or 80 or 90 feet in height and 10 or 11 or 12 feet in circumference and 200 or 300 or (once in a blue moon) 400 years, trees that felt the passing brush of wolf fur, that succumbed to beavers, trees that received the handed-down sense of lynx, bear, auroch, boar, reindeer, elk.

And dead trees. Dead trees standing, dead trees whose fall was broken and arrested by their neighbours so that they decay at an acute angle while their roots give up the ghost in mid-air, dead trees that lie prone but still give of themselves after death, feeding legions of bugs and birds and enriching the soil where they moulder. With the trees on one side and space on the other, you begin to sense the imminent change. Where there are trees they are no longer tall, they no longer walk the earth on elephant feet, they no longer come at you in clusters and snuff out the known world.

Then you reach the crest.

Then you reach the last tree.

You step beyond, and in that beyond lies one of the great revelations of this landscape or any other.

You have simply climbed above the forest, the forest that now tumbles away from you down to a hidden river, then flattens and sprawls then climbs again to hap the lower slopes of that hierarchy of all the Highlands' mountains – the northern Cairngorms in all their massive glory, thumb-printed by the ice into plateau-topped corries, sundered by the ice into the vee-shaped trench of the Lairig Ghru, a landscape of grander than grand gestures, of *huge* gestures, and from the very spot where you stand all the way to the mountains there has unfurled a magic carpet of every green shade in and beyond your imagination such that an early bard pausing where you now pause (but millennia before you) might have let a Celtic phrase escape from his lips that might be summarised and translated today as 'behold the Broad Plain of the Firs'.

You have not reached the last tree of course, merely the last one of the first part of the forest. The last tree is still miles away, and by then you will be breathlessly and hopelessly in the thrall of mountains. For the moment you have won both a respite from the trees and an insight into the nature not just of Rothiemurchus but also something of the nature of the Great Wood itself, for Rothiemurchus is its most tantalising echo. My response to arriving at this particular milestone is likely to be one of three recurring urges, depending on the day, depending on my mood, depending on the nature of the journey (a few hours out along the Gleann Einich track or a few days in the high country); sometimes two or three of those urges will surface simultaneously, and these will be the best of all days.

One: I want to sit with my back to that last pine tree and stare at all this for hours and make sense of it and write something of it down.

Two: I want to wander away off-piste, downhill, through the forest to the river, then on and on and up and up and make a necklace of the corries and finally breast the rim of the plateau where the known world expands into un-dreamed-of breadths and wraps itself in the sky.

Three: I want to devour more and more miles of this

track, knowing how the ocean of trees ebbs and flows and leans here and there towards the river, and the red deer and the roe stand and watch, and a tall and handsomely limbed pine bears strange fruit that gleam dully blue-black in the sunlight (a covey of blackcock), and a crested tit (a dapper little pinewood specialist with a crest like a crossword puzzle grid designed by Picasso) breezes across the track and lands upside-down on a pine frond six feet in front of me; knowing how the track wanders at last among the southernmost outliers of the forest, then abandons even these at the Tree of the Return, and that beyond there is such an amphitheatre of mountainsides with Loch Einich in its throat and that . . . and oh, what must it have been like 5,000 years ago before the people got hold of it all, when the only tracks were the work of such as deer and fox and badger and cat and the wolves that held sway over them all, over everything, and the anthem of their raised voices might travel five miles on a still evening . . .

*

And sometimes I get no further than the river and because I like to write where I am surrounded by my raw material I sit with my back to one more pine and unpack a lunch and a notebook and make a quiet space in my head and try and graft myself onto the tree and become a piece of the forest landscape and have the natives of the forest treat me accordingly. Sometimes nothing happens at all and I finish my lunch and the pages of the open notebook in my lap are punctuated only by crumbs. Sometimes, however . . .

. . . I had been watching a roe deer that had walked out of the trees on the far side of the river, watching her graze and scratch her ears with a hind foot, testing the wind, drifting almost soundlessly in and out of spotlights of sunshine, no more than a yard or two at a time. She was all curves and angles, poised, graceful, relaxed and alert at the same time. She was startled twice, once by a jay that screamed from the branches above her head, once by a raven I heard but never saw. She turned her back on me and began to browse her way back into the deep green of the trees and away from the river.

I watched the river instead. It was high and hoarse and fast, fuelled by a week of rain that had tumbled spring snow from the mountains, and it surged past within a foot of the top of its natural banks. Then a red squirrel came to drink. She was on the roe's side of the river and when she first appeared she was head-first and halfway down a pine trunk where she had paused, limbs splayed wide, tail flattened against the bark and pointing straight up the tree. It was a long stillness (for a red squirrel). What next?

She looked around. She watched the deer, which ignored her. She galvanised, scrambled down, bounced across a few yards of open ground to the water's edge then stopped dead. The water roared past a few inches beneath her feet. She leaned her nose towards it and pulled back, apparently fearful of the press of water. She sat back, tail curved now, arching up and then away from her back. She looked around. The idea that she was weighing her options was irresistible, but quite possibly wrong. But then she did this: She bounced back to the same tree she had descended, and hit the trunk running. The deer's head turned and watched, first with head inclined left, then right. Then it stepped away and missed the show. The squirrel took a right at the first stout limb and ran along it. She jumped. She landed on a much more slender branch, still running. I realised then that the slender branch belonged to a different tree. (I catch on more slowly than red squirrels in the matter of pinewood geography.) She reached the trunk of the new tree, descended a couple of feet, and (still running) took to another branch. This branch was broken about eight feet out from the trunk, broken but not snapped off, and beyond the break it angled down towards the water and rested not in the river but on a rock with a steep triangular face that was broken only where the river had worn out a little scoop. The squirrel ran out to the break, bounced over it onto the slope of the broken branch, reached the rock, spread herself wide and head-first across its triangular face and froze. Then she lapped all the water she needed from the scooped out hollow.

There followed a movement much too fast for my eye, in which she reversed her position so that she faced up the rock with her tail snug against her spine. Then she retraced her steps, back up the broken branch, over the break, along the branch to the trunk, up the trunk, along another branch, leaped over the oblivious roe deer's turned back onto the original tree, reached its trunk, ran up it and vanished into the canopy.

Questions:

Did she know about the broken branch?

If she did, had she used it before?

Did it come into her reckoning only when the river was too high and turbulent to drink from the bank?

Or was it the first time she had used it?

Did she weigh up other possibilities first?

Did she know that the rock held water?

If so, how did she know?

Had she seen it from above?

Was the whole thing a sophisticated, calculated man-oeuvre?

Or was it all an improvised, spontaneous work of instinct rather than intellect?

Answers: I have no idea.

As I walked back up from the river to the track, it occurred to me that moments like that one happen every day all over the forest, that while I find them astonishing and marvellous, in nature's scheme of things they are simply the punctuation marks that litter the workaday life of such a forest, that they have happened forever, and that every day that I am not in the forest to see them, they go on happening in their hundreds. There will have been a time when such a squirrel was hounded through the trees by goshawks, and on the ground by wolves and lynx, as well as the predators it is still familiar with today like pine marten and fox, and occasionally wildcat and more occasionally still, golden eagle. The goshawk, although it is very slowly edging north and west from strongholds in the Borders and Aberdeenshire, is still either rare or absent from almost all its former woodland strongholds.

About the golden eagle: where it still nests in trees (a very small percentage of the 400 or so breeding pairs in

Scotland, and almost all of them around the Cairngorms pinewoods) it will occasionally surprise a red squirrel, and you have to imagine that the red squirrel has a wariness of golden eagles within its DNA – they have, after all, cohabited here for a few thousand years. But I have seen red squirrel drays in the woody mass of an osprey eyrie, and an osprey eyrie looks a lot like a golden eagle eyrie, an osprey being basically an eagle-shaped bird. Even allowing for differences in size, wing shape, tail length, colour and voice, it suggests a sophisticated knowledge of bird species so that the squirrel is confident of its own safety in the company of the fish-eating osprey whereas the proximity of a golden eagle is more or less a death sentence. Such is the complexity of relationships among the natives of the Great Wood, such is the awareness demanded of the natives by such a world of trees.

*

Sometimes a single tree is as potent as a broad plain full of firs. Here is a case in point. The tree is lopsided. Here, I make an aesthetic judgment. The tree is perfectly balanced for its own requirements; there is no such thing as symmetry among Scots pines, and its south burgeons better than its north. You expect that – the hunger for sunlight in a hard climate. The lowest branches begin one third of the way up the trunk, leaning south. The north-facing branches begin halfway up the trunk and they are half the length of the lowest south-facing branches. The trunk is straight, but just as it reaches the crown it lifts a finger to the north, a slender counterweight to the south-facing bias of the foliage, a fingerpost pointed at the Arctic.

The tree seems to commend itself to me. Alone-ness may be part of the appeal, for alone has almost always been my preferred way through the landscapes of the Cairngorms. At a certain point on the track I start to watch for the tree, for it has become a kind of talisman. It has not always been alone. The corpses of dead comrades, each supplying its own bleached and still-rooted headstone, commemorate the conviviality of lost woodland. They stand all around or they lie where they fell. All are shrunken or broken or both. The notion of a battlefield is irresistible. But this one tree has survived alive and intact. It stands on a low ridge. Its situation confers status its modest size hardly merits, as if nature has made it a plinth to stand on. You see it from below as you walk the Gleann Einich track, but you do not see it against the sky. Instead nature has painted it into place on a mountain-shaped canvas, and it is that which accords it distinction in a forest of so many trees.

That mountain shape is the profile of Carn Elrig which is as solitary in the Cairngorms' scheme of things as the tree. Carn Elrig stands – uniquely – apart from the massif, severed by a moat of space as deep as the mountain. It is a lowly mountain by the standards of Cairn Gorm and Braigh Riabhach which it contemplates, but like the tree that keeps it company, it seems to summarise its own landscape – the Cairngorms in microcosm. It climbs from a thin frieze of trees around its north-facing prow through first heathery then bouldery phases to its own small, flat and gravelly plateau summit. It is as if nature had made a tiny scale model of what it had in mind for the Cairngorms, left it lying around for reference while it built the real thing, then forgot to tidy up after itself, leaving its prototype to its own fate.

One tree, one mountain, pine needles and granite. Be fruitful and multiply. For long enough, for those first 5,000 years after the great ice, and for God knows how long before it, they did just that, until the great grazing tribes of animals, fluctuations in climate, and the remorseless spread of the people began to unpick the Great Wood at the seams.

I like to walk until I set the tree against the mountain, the dark green against the dark grey (the white if it is winter), and by virtue of the changing sightline from the path below, move the tree until it aligns perfectly beneath the summit of the mountain, and make of that painterly composition a symbolic harmony of pine and granite and of the space between them that also binds them. I like what I have done with mountain and tree over the years, the way I have made them dance to a tune in my head. There is that point in the late afternoon or the evening, after a day up among the plateau spaces or Braigh Riabhach's corries, when the tree comes into view, but while I have been supping with the mountain gods it has wandered off on assignments of its own. See how far it has gone astray from the centre of the mountain! So I have come on it from a certain direction like a good collie penning a troublesome ewe, and I fix it with my collie's stare until it goes and stands where I want it. Then I admire their profound simplicity of form, their profound symbolic imagery. For tree and mountain are indivisible in my mind. Pine trees are as fundamental to the

Cairngorms as granite and wedges of old snow in August.

Each time I return to the Cairngorms, I head first for the Gleann Einich track and Carn Elrig. It has become something of a ritual. I walk with something of the purpose of the pilgrim, for with my every forward step I am trying in my mind to reach back. Uniquely among the woodlands of Highland Scotland, Rothiemurchus is far-flung enough, it is diverse enough in its repertoire of atmospheres and rich enough in the species of nature it sustains to be able to impose itself on a solitary wanderer seeking some kind of sense of the Great Wood of Caledon. Here and there, and for minutes at a time, I can convince myself that yes, this is what it must have been like.

It makes demands. The first is to slow my footfall. The second is that I must read nature. The third that I possess sensations. The first time, this was just my track to the mountains. But over the decades my mind has adapted to its force field so that now I think of it as its own world. Now we are forever asking questions of each other. 'The thing to be known grows with the knowing.' Now, within the first five minutes I am less of the outside world and more of the pinewood. I have become almost fluent in its slow speech. I walk to its pibroch rhythms. I possess many of its sensations.

It is at its best in sunlight, or in rain, or in a big wind, or in windless and heavily falling snow. If it is a sunny day in the world beyond the wood I see how the sun imbues the pervasive bottle green with other shades – black, dark blue, yellow, gold – and all of them still essentially green. Shadows are hard and brittle, not pliable and dappling like an oakwood. They lie among the trees like dark lace, each tree casting its shadows on several others like ragged netting. There are few pale shades in the depths of the pines and their companionable junipers. The air fills up with scent, a hybrid of pine and juniper. Nothing on earth builds a more fragrant campfire.

A rainy day intensifies the scent and layers it enticingly with bog myrtle blown in from unseen heathery clearings. A pinewood glows in rain; the very trees shine and their colour intensifies. The deeply contoured bark deepens to something improbably red. The junipers acquire a second fleece of silver.

On a day of boisterous winds the pinewood rocks. The canopy sounds like squads of jackdaws. The slenderest ends of trunks and limbs lean at unaccustomed angles and crackle and groan and squeal against each other so that you half suspect you have blundered across a treeful of capercaillies.

Heavy and windless snow cancels all of that, lays a pale grey shroud over everything that is dark and hard-edged and makes silence, utter silence that is somehow the opposite of peaceful – silence laced with tension. Possess that sensation too and imagine the impact of such a snowfall on hundreds of square miles of forest. Perhaps the Romans were finally unnerved by a forest made silent by weeks of heavy snow and lacked the stomach to wait for spring, knowing that the deeper the winter, the stronger the wolves grow. Perhaps the reputation of both forest and wolf forged something altogether more fearful and unnerving than the sum of the parts. The huge, slow aurochs felt it too, knowing that as the snow deepened and they weakened as feeding grew more difficult, the wolves were at their peak. The reindeer and the moose know it in today's Norway. In my head it is the pinewood's memory of wolves that puts the tension there when it snows hard and windless.

But I walk in today's wood and it is lit by that sun from the outside world and pungent with early spring. Walk softly, read nature, possess sensations. I scrutinise every small bird sound for the four-syllable signature of the crested tit. I stop dead at the first sound of it and I wait. There is a chance that sooner or later it will come close and a crested tit at close quarters is a pocket-sized piece of northness. This is their world. All they ask is a good tract of pine trees. They belong utterly in a wood like this. In Rothiemurchus (and in the other pinewoods of the Cairngorms: Abernethy, Glen More, Glen Feshie, Glen Derry, Glen Quoich, Glen Lui, Ballochbuie, Glen Tanar) they live in splendid isolation, nesting in dead pine trees, disputing territorial rights with tree creepers, disputes sometimes conducted in pairs. Such a dispute is a breathless dance of flat-out manoeuvres, blurring wings, tight turns, and loud irritation. I have never seen them actually come to blows but I have no doubt that it happens, in which case I would imagine that the odds are on the tree creepers, what with their outsize claws and sabre beak. Not all of nature's great conflicts over the 10,000 years of the Great Wood's story were fought out between wolves and aurochs.

The Great Wood of Caledon had many different densities of trees, from something like chaotic plantations – spaces crammed with young trees that had to compete with each other for the right to grow tall – to something like parkland. Rothiemurchus echoes that tendency, and suddenly something like parkland slows my pace again. Here is a new kind of space to contemplate, space punctuated by rounded giants of pines and stands of waving birch – a pale shade at last – then the first glimpse of mountains beyond – folds of huge, blunt, dark red shapes where the trees fail against their lowest slopes.

As soon as I leave the warm acres of the clearing, Rothiemurchus cools and closes in again, but from that moment on, I will walk in possession of that ultimate sensation in this landscape – the shape and the shade of the mountains' presence, sensed even when they are hidden from sight by the shape and the shade of the pines. Back on the track, Rothiemurchus settles once again into the familiar rhythm nature insists on, my own soft footfall and the elephant-slow progress of the wood as it slips quietly past, tree by tree by tree by tree. Then something new begins. The landscape begins to realign itself as the path begins to lean towards the river whose voice has begun to announce its approach well before it comes into view. And I begin to discern something particular and familiar in the lie of the land. And I know that soon a solitary pine tree will emerge from the horde, will detach itself from the press of thousands of others and stand there on a low ridge and the mountain shape of Carn Elrig will lift into the sky behind it. Something fundamental, something old and primitive slips into place and I begin to anticipate a particular set of landscape circumstances involving tree and mountain; I am navigating by the trees the way seamen used to navigate by the stars. And while it is true that no one needs to 'navigate' from Coylumbridge to Loch Einich today (the 'path' has been made to accommodate Land Rovers), there was a time when the earliest settlers here followed the paths laid down by the travellers of the Great Wood – red deer, roe deer, reindeer, fox, badger, lynx, bear, wolf – and they knew where they were when they could align two different features of the landscape into a particular configuration. So I walk, looking at the tree and the mountain, until I have aligned the tree perfectly beneath the summit of the mountain and made of that painterly composition a symbolic harmony of pine and granite and of the space between them that also binds them. I have made one of the oldest things that my species ever created as it began to explore the Great Wood – a landmark, a reference point. From here I know how long it will take me to reach the loch, how long to retrace my steps to Coylumbridge, how long to climb the mountain, how long to head out into the trackless west following the swirl of the upper limit of the trees until I reach the upper limit of the upper limit, the one true stretch of natural treeline in all Rothiemurchus, all the Highlands, all the Great Wood of Caledon. It is a kind of woodland shrine. Its name is spoken by foresters and silviculturalists everywhere with the reverence of the pilgrim: Creag Fiaclach.

CHAPTER NINE

Creag Fiaclach

Creag Fiaclach is an outcrop of Creag Dhubh, which is an outcrop of Sgoran Dubh Mor, which is the massively defining western flank of Gleann Einich, which, with a little help from Sgor Gaoith, Carn Ban Mor and the lower reaches of Braigh Riabhach, makes a box canyon of the glen and makes the shores of Loch Einich one of the most thrilling places in the land by day, and, by virtue of Sgoran Dubh Mor's miles of bulging cliffs, one of the uneasiest places I know to spend a night in a tent. On the other hand, to stand on the summit of Sgoran Dubh Mor just beyond the cliff tops is to feel as if you are under siege from the marshalled forces of sheer space that characterise the high Cairngorms. At the north end of Loch Einich you have left the trees four miles behind you. On the summit of Sgoran Dubh Mor they are about the same distance, which is much nearer than you might think and much higher too, for Creag Fiaclach is the setting for the uppermost gesture of the Great Wood of Caledon before it acknowledges its ultimate limitation and submits to the unstoppable power of mountain winds. Here is the only naturally occurring stretch of treeline anywhere in the land, and it concedes defeat at about 2,200 feet.

Imagine all the mountains of the Highlands that reach, say, at least 2,500 feet – all the Corbetts and all the Munros – and all the thousands of miles there must be that represent the 2,200-feet contour. Imagine too, for the sake of un-scientific argument, that half of these mountains once sustained significant native woodlands. Even if only half of that half was able to sustain trees at their natural upper limit, that's still at least a thousand miles of natural treeline. Yet the trees on Creag Fiaclach are all that are left of that treeline, about three to four hundred yards of it. If you would like some measure of the demise of Highland Scotland's native woodland, or even a metaphor for it, you need look no further.

You can reach Creag Fiaclach from the Gleann Einich track, gaining height gradually as you cut obliquely up through the contours by keeping just above today's treeline. Or you can wander the broad ridge north and downhill from Sgoran Dubh Mor to Creag Dhubh then bear north-west and still downhill until your gaze snags wide-eyed on the first trees, trees such as you have never dreamed of, but trees nevertheless. But better than either of these is to climb from the little showpiece of Rothiemurchus where the trees *are* the trees of dream. If the ambition of your expedition is to witness the last helpless shrug of the Great Wood as nature and only nature has shaped it and commanded it to evolve; if that is the be all and end all of your day's endeavour as it has so often been mine, then head for lovely Loch an Eilein, lovely but more or less sacrificed by the estate's mostly commendable efforts to reconcile the encouragement of tourism with landscape and wildlife conservation.

Start at Loch an Eilein because of the trees, because historically, even at the lowest ebb of Rothiemurchus's fortunes, the Grant family has accorded it affection and protection for purely aesthetic reasons. In the early years of the nineteenth century, for example, when timber extraction here was currency and the estate exploited it almost to the point of environmental catastrophe, Loch an Eilein was still not negotiable. Chris Smout writes in *Rothiemurchus – Nature and People on a Highland Estate, 1500–2000*: 'One small section round Loch an Eilein, Loch Gamhna and the Lochans was reserved and re-mained legally protected from sale because the family considered it an essential part of the "pleasure grounds of the manour of the Doune".' The 'manour of the Doune' is the historic home of the Grants of Rothiemurchus, and it was rescued and restored from the threefold ignominy of a hotel, a hunting lodge and a ruin by the present owner, John Grant.

Loch an Eilein's car park, visitor centre and walk-this-way footpath sometimes jar with the gorgeous setting, the 'honeypot' school of visitor management that I never much cared for, but the beauty of the place is undeniable and that beauty derives primarily from trees. So it is good to start here, where the pinewood is a douce and benevolent blanket, considering the nature of the day's destination.

Beyond the loch to the south is Coire Buidhe, a thing not so much of beauty as rarity in the context of a Highland landscape, being a corrie thronged with native trees. I simply love to climb up through the corrie, up through the trees, watching the slow transition unfold as the lush, springy and knee-deep understorey begins to grow sparse and shrinks to ankle-deep then vanishes among boulders, rocks, pebbles, gravel. The woodland unfolds a complementary transition as the dense, robust trees of the lower slopes begin to grow sparse and shrink in both height and girth. Then the corrie wall eases into the open mountainside, the wind hurtles down from the plateau, and with a quite startling suddenness (it startles every time) there is that weird Gulliver moment in which I find myself much taller than the trees.

That wind is quite capable of 100 miles per hour hereabouts, and up on the highest, widest reaches of the plateau in winter, 150 is quite feasible. I have just stepped out of the workaday world of a Highland pinewood into a land of quite uncompromising extremes, and it was all just the work of a few minutes, a few uphill yards. I gasp for breath, I look around at where I stand, and there are the crowns of the trees down there. For the trees' response to such a wind is to bow down to it, to flatten and grow sideways instead of upwards. I have never seen such trees other than in these few yards of Creag Fiaclach's frontier. Here, a juniper bush is as tall as a pine tree, 'tall' being a less than appropriate adjective for what is going on, for most are no more than waist-high and some no more than knee-high. Some are airy as thistledown, some as densely matted as a tabletop. All of them crouch and make what they can of the land's suddenly miserable store of nutrients by expanding outwards in every direction. I came this way once in low cloud with visibility down to about 50 yards. It was like wandering into a parallel and alien world populated by giant, sightless crabs. It is perhaps the most primitive place I have ever seen. The forces at work – the might of the wind and the puny tenacity of the trees – are things that pass my understanding.

Creag Fiaclach. Remember the name, for it is emblematic of a lost land, a broken frontier of the Great Wood. No, not a broken frontier, an overwhelmed one. The Romans may have turned back at the prospect of the Great Wood. Here is where the Great Wood turned back.

CHAPTER TEN

Glen Strathfarrar

The hills around Loch Monar were steep, orange and bare, and bedevilled by a mid-autumn wind with a mid-winter lash to its tongue. The orange of the thin, unnourishing grasses would fade to pale tawny inside a month. A dam hems in the loch and gives it that too-full look, commanding the River Farrar to leap rather than to seep into life.

I sat huddled over a sandwich, an apple and a flask of tea that cooled too quickly. I considered the long glen I had just travelled to arrive at this gaudy amphitheatre with its sullen, steel-grey watersheet fretting whitely at the dam wall. The hillside where I sat was sodden; its every gully and deer track gushed and muttered and oozed with water on the move. The glen that ended far to the east in the middle country of the Beauly River sprang from this blasted heath not 20 miles from the Atlantic seaboard.

Something fundamental was at work here; I had reached

an extremity of the empire of trees, a northern rampart of the Great Wood of Caledon. Wherever I have travelled among these remaining strongholds of the Great Wood, I have wondered if I might find such a final gesture sooner or later, and if I did, where I would find it and how I would recognise it. Would there be a last climactic landscape gesture with a spectacularly treeless northern panorama beyond, perhaps a sign saying 'here be dragons'? But instead, arguably the most beautiful treed glen of them all simply climbed and narrowed up and out of the realm of trees and lost itself in a huddle of bare, steep, orange hills, and there was a fitful drizzle adrift on a damnable north wind, a blink of sun that conspired wearily with the drizzle to make a pale rainbow, and there was a golden eagle that dallied in the high, cold air for 20 minutes, indifferent to all of it.

I don't believe hills like these were ever wooded. The ice that shaped them, whittling them like a dirk on a hazel wand, left no sustenance for trees. Bare rock bursts through the taut, orange skin of the autumn landscape. In hills where the Great Wood once flourished and was driven out, trees still cling to the steep gullies shunned by those tribes that grazed and felled and burned to excess; rowans and birches still spring from clefts in boulders and offer up green sprigs to nesting eagles. But not here. Not on this gaunt watershed beyond which the Highlands face west and contemplate the short, sharp descent to the ocean. These golden-eagle-tilted hills that crowd around Loch Monar felt the ice come and go, and they were as bare after it as they were before it, as they are now. The transformation after the long woodland miles of the glen below is fast and startling. Glen Strathfarrar, when you travel it from east to west, is a slowly evolving and increasingly intoxicating distillation of native trees in almost all their Highland moods (although without the breadth or scale of the Great Plain of the Firs). When such a phenomenon is heightened by all the shades of prime autumn, the headiness of it all is almost beyond words, which is a dangerous terrain for a nature writer.

Its douce beginning a few miles from the town of Beauly involves a short drive along a shyly signposted single-track road through a kind of coarse parkland, the comparatively wide and comparatively flat strath-like overture, complete with domesticated red deer. You suspect that the land and the deer will be wilder before the day is over. There is a locked gate by a cottage where, outwith the winter months, your car is issued with a (free) permit and you are asked (courteously) to be back by 6 p.m. If you choose to walk or cycle there are no restrictions – good luck, bon voyage and have a nice day. I swallowed my unease in the face of an estate road with a locked gate and a time limit; the good-humoured, hail-fellow-well-met welcome from the woman in the cottage and her eagerness to chat about the wildlife were mercifully mitigating factors. Also, the glen is a National Nature Reserve and I have never been one to quibble about restricting car access into the heart of one of these. But still . . . a locked road? In the twenty-first century?

Glen Strathfarrar begins to change character almost at once and plunges into a narrowing two-tone world of flame-shaded birches and glowing bottle-green Scots pines.

There will be countless variations on the theme of native trees in a Highland setting before the glen is done with you, but that twofold conspiracy of colours and tree shapes establishes at first glance both the character of the glen and your most durable memories of the time you spend in its company.

The River Farrar lays a dark and steely silver-grey thread through the weave of that tapestry-in-landscape, a thread enlivened by the too-whiteness of rapids. Trees crowd down to the banks where leafless alders greyly soften the pine-and-birch regime. But turn your back on the river and suddenly the slowly brightening mid-morning hillside birches swarm and smoulder, and spotlights of pale sun fan them into blazing thickets. The sorcery of trees has begun again.

The glen opens suddenly. The trees stand back and you see for the first time that they clothe the hillsides in climbing tiers, growing sparser as they climb. Low cloud is beginning to rise and fade, the sun beginning to cheer the land, and – also for the first time – you see distance, and in that distance too there are trees. The river has vanished. A small loch blinks in the growing brightness. I think of Thoreau at Walden: 'A lake is the landscape's most beautiful and expressive feature. It is earth's eye; looking into which the beholder measures the depth of his own nature.'

Our eyes meet – mine and the earth's. Mine are a northern blue, but this earth's eye is a kaleidoscope. The water by its farthest and nearest shores is milky white, but because the hour is windless and the surface utterly still, it reflects every shade of tree and hillside, but tones them down in one strenuous effort to contain them all. The darkest pines inverted in its darkest depths look almost black, and the most brazen birches are embers rather than flames. A glimpse of the hillside beyond the western shore is treeless and orange, and that too lays a broad band of a lesser shade clear across the earth's eye from north to south.

The southern shore is dominated by a particularly dense stretch of woodland, a persuasive emblem of a Great Wood, should you need such a thing. It creates the illusion of rising right in the middle of the loch to a regal crown of big pines. Some oak trees – the first I'd seen in the glen – are conspicuous in the throng, and again a grey fleece of alders wades into the water's edge, and again clusters of birch explode with yellow light among the dark pines. The earth's eye takes all that in and inverts it and makes of it one of the day's indelible memories, in fact one of the most indelible memories of my exploration of the entire realm of the Great Wood of Caledon. The lightest of showers drifts by and veils the trees, hissing softly on the water. It trails the hint of a rainbow and vanishes as if it has never been.

The northern shore of the loch is craggy, heathery, brackeny, and more sparsely treed. There are far fewer pines; more birches, scraps of holly and juniper, a handful of waterline alders, and here and there the showstoppers of autumn in such a glen – aspens. These flaunt a shade of gold that is quite beyond anything even the birches might dare. Beneath the crags the hillside is the dull red of faded heather and the rust of turning bracken, but between the trees and the water the hill grasses are the most vivid shade of orange. The loch accommodates all of these and repeats the hillside gullies in sweeping diagonal parallels. An hour in this kind of company drifts by in the blink of an eye while you measure the depth of your own nature. This particular earth eye is called Loch Beannacharan, which, unless my frail grasp of Gaelic misses by miles, means the Loch of Blessings. Amen to that.

*

I wonder about aspens. The aspen is one of those trees that seems to have got under the skin of the natives for a very long time, a bit like yew and hazel, although for different reasons. With the aspen, the whispering conspiracy of its constantly restless leaves was the cause of widespread unease. In old Scots the tree is the quaking ash, in Gaelic *critheann* or *critheach*, from *crith* meaning tremble, and its Latin classification is *Populus tremula*, the trembling poplar. In some parts of the Borders there is a local name, 'old wives' tongues' (the leaves are tongue-shaped, and they do gossip), and a more mysterious one in Ardgour – *sron a crithean* – trembling nose.

The implication in the folklore of many countries where aspen grows is that there is a sinister aspect to the restlessness of the leaves. Alexander Edward Holden wrote in his book *Plant Life in the Scottish Highlands* (1952) that 'the old Highlanders believed this was so because the Cross of Calvary was made from aspen wood and for this reason its leaves can never rest'. In Glen Strathfarrar on a high autumn afternoon I would have thought that old Highlanders with religious axes to grind might have found that the imagery of the burning bush might have better suited their cause than Calvary. In any case, and with the significant qualification that the nearest I have ever been to the Holy Land is Switzerland, I understand it to be a hot, dry and dusty place (the Holy Land, not Switzerland), whereas the aspen tree's extensive shallow roots send up suckers to create small thickets, a technique that requires cool, damp conditions. I doubt if an aspen tree ever grew in the Holy Land, and if it did it was so rare and shrunken that it would never have been the material of choice for the local carpenter when he was commissioned to make three crosses.

I was about to embark on a piece of clinching research on the subject of trees in the Holy Land when I discovered that Alistair Scott had already done it in *A Pleasure in Scottish Trees,* and that his research propped up my best guess: no aspens, ever. He also observed that 'aspen needs a fertile mineral soil which is one reason why you will often see it by a burnside', and now that I think about it, so you do. He also wrote that 'frequently it will be a clump deriving from root suckers. The individual stems are short lived; the clump is, I suppose, immortal.' Interesting choice of words in the circumstances.

Yet the Calvary tradition endures. The ancient source of it is lost but its modern manifestation stems from a much-quoted passage in Alexander Carmichael's nineteenth-century *Carmina Gadelica*: 'Clods and stones and other missiles, as well as curses, are hurled at the aspen by the people of Uist because it was used to make the Cross on which Christ was crucified. No crofter or fisherman would use aspen wood for any purpose.'

The *Carmina Gadelica*, a collection of Gaelic hymns, poems and other utterances, ran to six bilingual volumes. Many entries referred to the place of the native flora in the Gaelic oral tradition. A hundred years after his death there is still controversy about just how much Carmichael tampered with the fruits of his huge research. But why Uist of all places was so militantly anti-aspen is not explained. Scott adds a telling aside that the aspen grows well in the Western Isles. Maybe the islanders' aim with the clods and stones and other missiles was indifferent, likewise the potency of the cursing.

Alistair Scott's most intriguing revelation is rather more reliable, and follows late twentieth-century research that 'holed below the waterline' the belief that seed production of Scottish aspens was very rare and that its presence could only be expanded by propagation: 'It is now clear that aspen does produce seed from time to time, more in some places than in others. At long intervals, after very warm summers like 1995, it can produce spectacularly large quantities. The seed is as light as a dandelion seed and will be blown as far. It has a short natural life and will only germinate and grow if it falls on a suitable substrate, where competition is sparse or absent. Here is now a satisfactory explanation of why aspen was so successful in the early post-glacial.'

What Scott did not explain is why there are no big aspen woods in the landscape of the Great Wood. There are expansive miles of aspen woods in North America where the natives drive all day in the fall just to stand and stare at the spectacle of trees aflame. The aspens and the maples are the superstars of the show, and the people hurl nothing more ominous at the aspens than admiring glances. They call it 'leaf-peeping'.

A few years ago I was in Alaska to make two radio programmes for the BBC's Natural History Unit with producer Grant Sonnex. We stayed a couple of nights at an inn near Gustavus in the Alaskan south-east – an inn in an aspen wood. We had got my voice piece down on tape and Grant wanted to record the ravens that haunted the woods and whose voices were the unending anthem of that place. But the comings and goings of float planes, dogs and pick-up engines were too intrusive. So he decided to get up at first light before the people were up and about. His alarm went at 5 a.m. He woke to the sound of heavy rain – and the unadulterated cries of ravens. He liked that combination so he dressed from head to toe in waterproofs, protected his recording equipment as well as he could, and stepped out of the door – into a morning of quite breathtaking beauty lit by early morning sun. What he had thought from inside the building to be the sound of heavy rain proved to be a vigorous breeze blowing through the leaves of thousands of aspen trees. The characteristic leaf-whisper of a Highland copse was magnified to something like a hoarse roar – a soft, hoarse roar – by the sheer numbers of the trees. But he made a beautiful recording of ravens, laved by the windsong of aspens. When he told me the story over breakfast a couple of hours later, there was still an edge of wonder in his voice.

One way or another, the aspen tree has never lost the power to impress susceptible minds. Twelve years after Alaska, where the radio programmes were about the re-lationship between people and wilderness, I startled myself out of a kind of spellbound trance induced by a copse of honey-gold riverbank autumn aspens in Glen Strathfarrar, and what startled me was the sudden memory of Grant in Gustavus telling me how he stepped out of the inn on a late-August morning just as the sun hit the aspens, a morning he greeted swathed in Santa-Claus-red Goretex. I was in Glen Strathfarrar to write a book about the Great Wood of Caledon, or to put it another way, the relationship between people and wilderness. Now that I think about it, my response to both events was identical; it was to wonder why – and to regret – there are no big aspen woods within the Great Wood.

In Glen Strathfarrar, the trees relented briefly and the view widened over a flat and grassy floodplain where hordes of mallard and wigeon grazed around the great black feet of eight whooper swans (five cygnets, their parents, and an adult hanger-on) and two very strange looking swans indeed. The woman who had unlocked the gate had mentioned the swans, and in response to my eager inquiry (swans have had a hold on my writing and my imagination for half my life) she imparted exciting information. In the normal course of events, whoopers migrate from Iceland for the winter, and usually arrive in Scotland in October and November. I asked when they had turned up. Oh, they don't leave, she had said.

It seems one of the adults had been injured a few years before, and been unable to make the spring migration home, so its mate stayed too. They nested in the glen instead of in Iceland, and they had been there ever since. I have encountered similar situations before, and there is a handful of nesting whooper swans scattered across the north of the country, some for just a year or two, for the migrant urge is the strongest force in the natural world, but others fight it for various reasons and try and settle. What was odd about this little gathering was the two extra swans that kept the whoopers company. They looked like mute swans except that their beaks were neither one thing nor the other, as if someone had botched a repair job, run out of mute swan parts and used whooper parts instead. They were hybrids. It seems that one of the whoopers had mated with one of the local mute swans, and these were the result, the very rare offspring of a very rare set of circumstances, nature catching itself out and correcting its own mistake by producing creatures that could not reproduce, thereby halting the hybrid line in its tracks.

It is possible that in colder eras than this one, when Iceland was altogether too icebound to offer summer sustenance to birds like these, the whooper was a resident in the Great Wood, and that what I was watching was a glimpse of the clock rewinding. Bring on the aurochs, the bears, the lynx, the beavers, the wild boars, the wolves.

Glen Strathfarrar's skinny little locked road climbs purposefully towards the dam, and the trees begin to fall back and thin out. I sensed finality. The end of something was in the air. But there was to be one last gesture, a final unforgettable flourish of the Great Wood, a crowning glory, a somehow symbolic arrangement of the glen's landscape elements into a distillation of all of them, and designed to linger long in the earth-eye of the beholder. Just where the river emerges from a dark and rocky gorge below the dam, and wheels away south into a sudden glimpse of sunlight, a cluster of about 20 big Scots pines thickens a corner of the hillside, and in the ancient tradition of Highland pinewoods sends a few outlier trees capering away up the skyline. A handful of birches clings to the shady slope beneath but they are poor trees struggling to hold their own. Yet the pines thrive, even here, for there is no soil too poor, too thin, too acidic for them; no terrain is too rocky or too steep. The Scots pine's success in the Highland landscape is the consequence of a perfect and very primitive design. The species is much, much older than many of the native trees that clothe our landscape today. It is sustained by a root system that both powers deep underground to find moisture on such well drained land and also sprawls horizontally amid the humus of the surface of the land. So it is well anchored too.

My 'earth eye' (I have taken Thoreau to heart, as you see) still beholds that small gathering of trees, still remembers their natural grace, their elegant stances – straight-trunked, broad-limbed, wide-crowned. I remember too how the land seemed to enfold them in a kind of possessiveness; how they dominated the whole landscape view from the middle ground; how a single, distant, sharply etched mountain summit was somehow ennobled by them.

Beyond the gorge the road crosses the dam and dwindles to a dead-end. Then, after a short walk among the hills surrounding Loch Monar, sitting hunched over a sandwich, an apple and a flask of tea that cooled too quickly, I thought of that group of trees as a signature. The Great Wood had signed off.

CHAPTER ELEVEN

The Great Woods

The people were few and they clung to the coast. They had come in boats, creeping up the outermost edge of Europe. Five thousand years ago, the land we now call Scotland was still hinged to the landmass of Europe by a land bridge, but even so they saw it as an outpost of that edge. Its every coastal mile glowered at them, bared its wooden teeth at them, for the trees grew down to the high tide line and seemed from the sea to swarm over the low hills and halfway up the highest mountains. The trees were pine and oak, birch and hazel, willow and rowan, juniper and alder, aspen and ash. The people understood wavefall and whale, not woodland and wolf.

There seemed no end to the trees. But it was an illusion, a seafarer's misconception of the lie of the land. In Scotland of all places, both geographically and historically and now as never before, there has always been an end to the trees.

159

This land was never jungle. Trees never covered the land. Even if the land was capable of sustaining trees in its every acre which it palpably is not – the many grazing tribes that have thrived here needed grassland and wetland, and the tree-felling tribes like ourselves and the beavers began making inroads almost as soon as there was a forest to fell.

Nevertheless the reputation of an infernally hostile Great Wood was whispered among seafarers and grew from there, probably over several thousand years. There were great firths we know now as Tay, Forth, Clyde, and these gave access to gentler lands, fertile ground on the floodplains and a low-lying, eastern coastal plain, woodlands that did not daunt. But in the Highlands of the north and the north-west all was hostile and defended by a Great Wood. They may have used those very words or something like them in their various tongues, and these sprang from mouth to mouth, crew to crew, expedition to expedition, country to country, and the legend of the Great Wood was born. When you walk through the landscape of the 21st-century Highlands, you walk in the footsteps of that legend. The remnant woods are its footprints.

In some ways, the Great Wood is like the wolf. Its reputation was forged by generation after generation of storytellers, and it was rendered all but extinct by people who believed the legend. That reputation is like a cairn of stones, and each generation added its own armful of stones to the cairn. And the reality was buried under a mountain of stones and lost in the fogs of the gathered millennia. Yet historians, archaeologists, biologists and silviculturists are all more or less agreed now that there was a time about 5,000 years ago when the native woodlands of Scotland reached their greatest extent. It is reasonable to assume then that given the general difficulty of the Highland terrain, especially a Highland terrain without roads, that any extensive tract of Highland forest would be especially troublesome to travellers. I had begun this book with the idea that I might try and pin down what such a forest might have amounted to, and for a while I was thinking of a single forest entity. But by threading together the remnants of native woodlands, by looking again and again at the lie of the land, by cashing in on the accumulated detritus of a lifetime of watching and wondering about the ways of nature in such a landscape as Highland Scotland, and especially by remembered conversations with people whose judgment I trust and by consulting the published works of people like T.C. Smout, Alan R. MacDonald and Fiona Watson (*A History of the Native Woodlands of Scotland*), Alistair Scott (*A Pleasure in Scottish Trees*), Don and Bridget MacCaskill (*Wild Endeavour, Listen to the Trees*) . . . after all that I have come to a different conclusion of my own. I think there never was a Great Wood of Caledon. I think there were four Great Woods.

Their characters are so different, their geographical contexts are so self-contained, and each is bulwarked from the others by the kind of mountain barriers that preclude any likelihood that they could ever have amounted to the constituent parts of a greater whole. Their boundaries may well have ebbed and flowed, for trees rise and fall like tides in the course of two or three thousand years . . . for if the Great Wood is anything at all it is a continuum. But there was constancy at their core. So what makes sense to me are these four woods:

One – The Trossachs, which name I use under protest and for convenience only, for since the Loch Lomond and the Trossachs National Park was invented it has become lazy shorthand for everything from the east shore of Loch Lomond in the west to the east end of Loch Earn in the east, and from the Lake of Menteith in the south to Balquhidder in the north. With the shining example of Glen Finglas at its heart and good oakwoods elsewhere both within and out-with the Forestry Commission's large landholding, its prospects are reasonably bright.

Two – Glen Orchy and Rannoch are, historically, the greatest tract of the Great Wood, or as I have come to think of it, the greatest of the Great Woods, from the north shore of Loch Awe in the west to Loch Tay in the east, and from Glen Dochart and Strath Fillan in the south to the Black Mount, Rannoch Moor and the Black Wood of Rannoch in the north. Its scope was immense: the broad trough of Glen Dochart, the mountain-climbing pines at Strath Fillan, the teeming trees of Glen Orchy and Glen Strae, the great spaciousness of Loch Tulla, Loch Ba and the lightly wooded Rannoch Moor, the dense Black Wood, the water-highways of Loch Tay and the Rannoch-Tummel chain; its flora and fauna must have been among the richest and most diverse in the land. Its potential for 21st-century conservation is limitless: wildlife reintroduction up to and including the wolf (and the beaver is a natural in such a land of woodland and water); the rehabilitation of much of the historic woodland, centred on the grand gesture of replanting Rannoch Moor, and diversifying the species and lightening the commercial burden in Glen Orchy; expanding the oakwoods of Glen Dochart, the pines in Strath Fillan, and enhancing the mix of native woods around Loch Tay (the Bolfracks estate on the south shore near Kenmore is already setting an enlightened example). If Scotland was ever minded to create and buy a national park to be owned by the nation and to manage it for nature, it should be this one.

Three – The Cairngorms, bounded by the Dee and the Spey. Years ago, in a book about the Cairngorms called *A High and Lonely Place*, I wrote that 'there is a uniformity of presence which attends every step of the way from Spey to Dee, and it is that uniformity, translated by Seton Gordon as "the spirit of the high and lonely places", that so distinguishes and dignifies all the Cairngorms, all their moods, all their heights and depths, all their landscapes.' I have never felt any need to revise that assessment. The native woods, especially Scots pine and birch, with good dollops of juniper and aspen, define the region as naturally as do the mountains, and the trees can be persuaded to encircle the massif as they once did, and to reclaim a natural treeline at somewhere around 2,000 feet. There is nothing else in the land like the Cairngorms pinewoods, and admittedly rather late in the day we have begun to recognise their worth and to begin at least a piecemeal restoration. But the only worthwhile task is, in the words of the Woodland Trust project's ambitions for Glen Finglas 'to restore native woodland across its full natural range'. In the wider Cairngorms, that is a mighty task, but it is a mighty landscape and it deserves only our mightiest efforts.

Four – The West Highlands, from Glen Strathfarrar in the north and the other parallel glens west of the Great Glen, and as far south as the Sound of Mull; and I have no difficulty making a case for extending the Sunart Oakwoods project across the Sound and onto Mull itself, as the trees did when the Great Woods were at the height of their greatness 5,000 years ago. This is our oceanic wood, subject throughout its length to a climate engendered by proximity to the Atlantic seaboard and the Gulf Stream, and hung on a skeleton of long and parallel west-running glens, lochs and sea lochs. Its strongholds are at the north and south, but throughout its length there are fascinating survivals and hints of ancient woods – from the pines at Knoydart, where the John Muir Trust is harvesting local seed to restore lost woodland (the character of these trees is quite different from the pinewoods of the Cairngorms and the seeds of one will not flourish in the landscape of the other), to coastal scraps of hazel woods that are among the least disturbed patches of woodland anywhere, and may be the only places left where we can look the original wildwood in the eye.

So that, I believe, is the true nature of a historic forest that has come down to us through the ages as the Great Wood of Caledon. I believe too that the long human occupation of the landscape that predated the Romans is much likelier to have invented and sustained the notion of a Great Wood than Tacitus and Ptolemy. I think it is more likely that the nature of legend suited the Romans' purpose when their nerve finally failed them, and they turned their back on the unconquered forest of legend. And because it was in their nature, they wrote it down, and it may be that the first formal recognition of the idea of a Great Wood of Caledon arose then. But by then the wood that they encountered had been greatly reduced from that high point 3,000 years earlier by natural events – climate change, flood, fire, and mysteries we have yet to solve, such as the drastic decline in Highland Scots pine across its whole range around 2000 BC. Why just pine? Why so widespread? In any case, the Great Wood of Caledon was not half the wood it had once been, and the Romans were going anyway.

But these Great Woods, or at least what is left of them, have a future as well as a past. What are we to make of them, we who profess to understand the concept of conservation, we who support conservation charities in our millions, we who have invented national parks, national forest parks, sites of special scientific interest, national nature reserves? Must we do anything at all? Well, strictly speaking, no. Ours is a wooded country. Trees will grow if we let them, so why formalise the process to make some kind of accommodation with an ancient notion of a Great Wood of Caledon that never existed in the first place?

There again, it was actually quite a good notion, and even if it was only ever partly true at best, and only in some parts of the country, and thousands of years ago at that, it is an idea that is worth trying to live up to in the twenty-first century. Biodiversity is a new idea, or at least it is new to people, no more than a hundred years old, and perhaps John Muir was among the first to give it a voice. But if you can come to terms with what I might call nature's mind, biodiversity is actually the oldest idea on earth, the one that makes sense of everything around us. ('I believe in God but

I call it nature.') We now profess to embrace it as a species. If we are serious about advancing its cause then the thing to do is plant trees. If we increase tree cover we also increase and diversify habitats, and therefore species. A native forest, once it is established, regenerates itself and expands to fill the land available to it. It matches tree species with soil conditions and determines the density of tree growth. Birds, beasts, plants and insects follow. Unless they are being reintroduced following extinction you don't need to put them there. They just turn up. Word gets around. A few years ago I spent four days camping on Pabbay, a couple of links down the chain of the Western Isles from Barra. Back in the 1870s a huge sand-blow had inundated about a third of the island. Now, two thirds of the island is rock and bog and rough grass, and the other third is deep sand and marram grass. In spring and early summer the sandy end is drenched both in primroses so thick that you can hardly walk there without treading on blooms, and in the song of dozens and dozens of skylarks. No one told them about the sand-blow, about the sudden change in conditions on a tiny island in the Atlantic. They just turned up, and every year the primroses still bloom and the larks still sing. If it can happen there, it can happen in four great mainland woods where it is in our gift to make them better woods, and more hospitable to more tribes of nature.

Besides, we owe a debt in this regard. From the moment we began to walk in our post-glacial land about 9,500 years ago, we became accustomed to taking from the land, to exploiting nature. Over the millennia we have become extraordinarily inventive in this particular endeavour.

Native woodland is infinitely renewable, but only if we let it be, only if we are prepared to give it space, only if we are now in a frame of mind to give back to nature something of that which we took away. It is happening now in a small way, but the pace of change and the scale of change are a long way short of significant. What we do have is hundreds of square miles of spruce-dominated forest that is routinely clear-felled and replanted with an industrial mentality whose effect on the ground is often loathsome to both people and nature. Yet a simple change of philosophy, a rethink of the national forest strategy and the role of the Forestry Commission within it, could quickly begin to transform the landscape while the restoration of native woodland proceeds at its own essentially slow pace. The Scottish government has a policy of increasing Scotland's woodland cover to 25 per cent from something like 17 per cent at present. It is a laudable ambition, and it may be argued that the target is too modest, that a third rather than a quarter of the land should be wooded – but only if it is the right kind of wood. Working within those four Great Woods the opportunities exist now to consolidate, conserve and expand the purely native elements within them, but also to blend well designed commercial plantations with native trees, natural spaces, grassy clearings, standing water, wetland, and open hillsides to create a new kind of Great Wood, one that reflects the realities of the tree species available in Scotland in our own time. We will never eradicate non-native trees from the entire country, nor should we waste our time trying, although there are worthwhile local endeavours like Glen Finglas where they have been removed to make room for native woodland. And nor should they dominate so many Highland glens the way they do.

Alistair Scott had some wise words on the subject:

Scotland has one of the best climates on earth outside the tropics for growing a wide range of trees. In essence this is the Gulf Stream effect. Scotsmen roamed the world throughout the last two centuries sending home seeds and plants from everywhere. The consequence is that there are, growing happily somewhere in Scotland, well over 1,000 introduced tree species.

The native tree flora conspicuously lacks, of the world's great timber trees, a spruce, a fir, a larch, a beech and a maple. We have gained, or should have gained immeasurably from the arrival here of Sitka spruce, Douglas fir, European larch, common beech and sycamore. Between them and our native timbers we have a better resource base than anywhere in northern Europe and as good as any in central Europe . . . Which is not to argue that the effects of these arrivals on our landscape have been everywhere benign. Sitka spruce was planted in too many places where no trees should be. It was, and to some extent still is, too often cultivated like a field of wheat, not as a complex forest. Sycamore is usually best kept out of nature reserves. And so on. But we have only had a century or two in which to experience these new trees and the lessons have already been learnt, or most of them have . . .

There are many good reasons to plant more trees and to take the time and trouble to do it well. They create the most benevolent of all nature's habitats. They create opportunities for a greater diversity of wildlife species of all kinds. They create a counterbalance to greenhouse gases. They create stable and long-term rural employment for people, because managing large forests both for nature and for timber is labour-intensive work. If local people are employed they create the circumstances that bond people closer to their place on the map, giving them a stake in the environment of that place. They create opportunities for recreation in a beautiful environment. And they matter for their own sake. And they should matter to us because of the debt we owe.

There is a place where I walk often, a walk that begins on a forest road between Strathyre and the north end of the west shore of Loch Lubnaig. There is a high point on that road at which the loch first comes into view, and it stops me in my tracks every time. I have met many people there, both locals and strangers, and not one has ever been indifferent to it. The foreground is steep, rough grazing that falls away to lower, wetter ground. It is lightly wooded with birch, juniper, hazel, oak, and as it descends the tree cover yields to willow and alder, and an area of grassy wetland becomes a substantial lochan. Beyond the lochan, the River Balvaig heads for the loch by way of something like an avenue of alders. It enters the loch between two bays thickened with tall grass and reed beds. There are oakwoods reaching far down the west shore of the loch, and above them hillsides of mostly Sitka spruce and larch, some birch. The planting design has a good, open feel. On the far side of the loch there is more plantation forestry with pines and larch let into the predominant spruce, and a distant skyline where the loch vanishes round a bend is one of tall firs. The two water-sheets – the lochan and the loch – bind the whole view together. In winter especially, it looks as if you could be in Norway or southern Alaska.

There are two difficulties with what unfolds here. One is that the commercial part of the equation is too dominant. The other is that although care has been taken with the planting design, every part of both hillsides will eventually be clear-felled rather than thinned, and this in the very heart of a national park.

So there is a philosophical question to be addressed, and it is whether or not we should demand of the Forestry Commission, of the national park, of the Scottish government, that the interests of landscape and nature should take precedence over the interests of the timber industry. I think it should, but then I am a nature writer rather than a forester, and you would expect me to say that. But I also think that a forestry industry that was compelled to put the wellbeing of the landscape first would breed better foresters whose jobs were more rewarding, and that a more diverse forest would create a greater diversity of timber species as well as of natural habitats. There should be a great deal more to our timber-growing industry than an endless conveyer belt of Sitka spruces that mostly gets pulped. As Alistair Scott wrote, 'How satisfactory, if we could reconnect the trees growing out there with the wood that pleases us in here.'

So here is a work for the twenty-first century. I think there were four Great Woods, and it may be that there was a time when they were all connected into a single entity, but if it did happen it was for a very short time, and it was long before the Romans knew where we lived. The four Great Woods should be the focal points of a new era for our native woods and for the kind of multi-cultural woods that are most likely to constitute the future of this portion of the planet. We may well argue we have got the balance right the day a few human generations hence when our descendants can point to a particular feature of a miles-wide Scots pinewood where a stupendous symbolic grove of Sitka spruce bursts through the canopy and soars 150 feet into the cool Highland air. And the example set by these four great national woods will rub off so that we create new community woods everywhere.

Meanwhile, at Fortingall, there is a stone wall that precludes the possibility of a new generation of yew trees reaching out from what is arguably the most important growing thing on earth, and I would like it symbolically removed. We owe it to nature and to all the trees that ever set foot on Highland soil to ensure that nature's idea of succession can take place, rather than our own idea of it. We owe it to the Great Wood that was, whatever it was. We owe it to the Great Woods that can still be. And we owe it to the one tree in all the land that may well be truly immortal.

Out of the Trees
(From the painting by Rob MacLaurin,
Gallery of Modern Art, Edinburgh)

We were long days in the trees,
Sun-snuffed, song-starved.
Eagle, owl, bear-prowl, wolf howl –
these scored the harmony
of what lived easily there,
though they noticed us and passed by,
fellow-travellers as tree-girt as we.

Yet we learned that primitive tribal fear
we thought lost beyond recall, but known
to explorers whose idea of 'map'
was carried in their heads: they'd lap
from rivers (like bear, like wolf)
until the day they thought
to cup their hands.

One evening the trees began
to end: colour, sun, song began
to tend us from beyond,
and we saw the yellow mountain
and the tribal-ness shrank
with the shadows and we drank
boiled tea from cups.